Spring Boot 2 攻略

[英] 马特·戴尼姆(Marten Deinum) 著
张楚雄 张 琦 译

清华大学出版社
北京

Spring Boot 2 Recipes: A Problem-Solution Approach
Marten Deinum
EISBN: 978-1-4842-3963-6

Original English language edition published by Apress Media. Copyright © 2018 by Apress Media. Simplified Chinese-Language edition copyright © 2019 by Tsinghua University Press. All rights reserved.

本书中文简体字版由 Apress 出版公司授权清华大学出版社出版。未经出版者书面许可，不得以任何方式复制或抄袭本书内容。

北京市版权局著作权合同登记号　图字：01-2019-5887

本书封面贴有清华大学出版社防伪标签，无标签者不得销售。
版权所有，侵权必究。侵权举报电话：010-62782989　13701121933

图书在版编目(CIP)数据

Spring Boot 2 攻略 /(英)马特·戴尼姆著；张楚雄，张琦 译. 一北京：清华大学出版社，2019
书名原文：Spring Boot 2 Recipes: A Problem-Solution Approach
ISBN 978-7-302-53949-0

Ⅰ.①S… Ⅱ.①马… ②张… ③张… Ⅲ.①JAVA 语言—程序设计 Ⅳ.①TP312.8

中国版本图书馆 CIP 数据核字(2019)第 224326 号

责任编辑：王　军　于　平
装帧设计：孔祥峰
责任校对：牛艳敏
责任印制：李红英

出版发行：清华大学出版社
　　　　　网　　址：http://www.tup.com.cn，http://www.wqbook.com
　　　　　地　　址：北京清华大学学研大厦 A 座　　　邮　编：100084
　　　　　社 总 机：010-62770175　　　　　　　　　邮　购：010-62786544
　　　　　投稿与读者服务：010-62776969，c-service@tup.tsinghua.edu.cn
　　　　　质 量 反 馈：010-62772015，zhiliang@tup.tsinghua.edu.cn
印 装 者：三河市君旺印务有限公司
经　　销：全国新华书店
开　　本：170mm×240mm　　　　　印　张：18.5　　　　　字　数：427 千字
版　　次：2019 年 11 月第 1 版　　　印　次：2019 年 11 月第 1 次印刷
定　　价：79.80 元

产品编号：083283-01

推荐者序

Spring Boot 框架作为当前最炙手可热的服务端开发框架之一，已被互联网业界的开发者拥趸为第一首选的平台。本人从事服务端开发十多年，从原始的 Web CGI、Servlet、EJB、Spring 开始，服务端的开发随着框架的完善，越来越便捷。尤其是 Spring Boot 的横空出世，开发者只需要使用简单注解和配置文件就能轻松搭建起一个企业级应用，可将更多时间和精力都集中在业务规则的实现上。本书的作者 Marten Deinum 作为开源 Spring Framework 的贡献者，对 Spring Framework 理解深入。当然，大家也不必担心，此书并不是单调的理论描述，更多是从实战及实用的角度出发，深入浅出地把读者带入 Spring Boot 的奇妙世界。本书的译者张楚雄技术翻译功底深厚，用中文原汁原味地将原书的精髓呈现给了读者，本书的确是一本既适合初学者学习又适合资深工程师参考的 Spring Boot 专业好书。

咪咕互动娱乐有限公司高级系统分析专家 潘伟

Spring Framework 是 J2EE 技术体系中的一套关键性开发框架，所有 Spring 模块的核心理念是依赖注入和 IoC 控制反转，这两个特性可以帮助程序员轻松地开发出高内聚、低耦合的应用程序，是面向对象、面向服务编程的有力保障。

但在开发 Spring 应用程序时需要进行很多配置，例如在使用 Spring MVC 时，需要配置组件扫描(Component Scan)、Dispatcher Servlet、视图解析器(View Resolver)、Web Jar(用于提供静态内容)等。为了精炼开发过程，Spring 家族中的 Spring Boot 横空出世，Spring Boot 首先会根据应用的 CLASSPATH 已存在的配置，自动装配应用程序框架所需要的各种基本配置，其次它提供了 Web、JPA、Rest、WebService、Test、Security 等常用的 Starter 模板，模板自动装备了各种依赖组件，最后它还具备运行监控、日志跟踪等可视化运维功能，这一切极大降低了应用开发和运维的难度。目前，Spring Boot 已经成为 Spring Cloud 等微服务框架的首选基础支撑技术，是软件从业人员的必备技能之一。

本书深入浅出地全面介绍了 Spring Boot 技术的前世今生，不仅从技术代码角度，更从架构设计角度讨论了 Spring Boot 的一些深层次思考。强烈建议读者仔细研读本书，并结合自身业务领域，举一反三，相信必能受益良多！

南京天溯自动化控制系统有限公司资深软件架构师　侯逸文

通过使用 Spring 框架，现代程序员可以比较轻松地开发 Web 应用。然而在 Spring 框架的搭建过程中，人们往往需要重复性地进行代码复制及配置文件的撰写，这个过程枯燥乏味且容易出错。科技因懒而进步，通过自动化配置，Spring Boot 大大简化了项目的搭建成本，让开发者更加专注于业务逻辑。本书通过大量代码示例和手把手的教程，能够让你充分利用 Spring Boot 的巨大潜能提升研发效率。行甚于言，本书将指导你编写出更优质的代码！

华泰证券公司 DevOps 产品经理　李青

作者简介

Marten Deinum 是开源项目 Spring Framework 的贡献者，也是 Conspect 公司的 Java 技术顾问。他为许多小型和大型公司开发和设计软件，主要使用 Java 技术。他是一个热情的开源用户，也是 Spring Framework 项目长期的追随者、用户和倡导者。他担任多个职位，包括软件工程师、开发主管、教练、Java 培训师和 Spring 培训师。

技术审校者简介

Manuel Jordan Elera 是一位自学成才的开发人员和研究员，他喜欢学习新技术，进行实验，并创造新的集成。Manuel 获得过 Springy Award Community Champion 和 Spring Champion 2013。在空闲时，他喜欢阅读《圣经》，用吉他作曲。Manuel 被称为庞贝博士。他曾为 Apress 审阅过许多书籍，包括 *Pro Spring, 4th Edition* (2014)、*Practical Spring LDAP* (2013)、*Pro JPA 2, Second Edition* (2013)和 *Pro Spring Security* (2013)。

致　　谢

尽管这是我编写的第四本书，但我仍然对创作这本书的工作量感到惊讶。不仅是我编写内容花了很多时间，Apress 出版社的非常敬业的编辑也花费了大量时间。从编写我的第一本书开始意识到写一本书是很困难的，然而，写一本关于尖端技术(Spring Boot 2.1)的书就更难了。为预发布/测试版的软件编写书籍需要重写代码示例和相关的内容。这有时会让我和我的编辑经理 Mark Powers 发疯。所以很抱歉我不得不重新编写很多内容，总的来说，这比我最初想象的要花更多的时间。虽然历经种种艰难，最终还是完成了这本使用 Java 11 并涵盖 Spring Boot 2.1 的新书。

我要感谢 Apress 出版社的所有编辑，他们在保持图书质量和内容的同时，努力让我按时完成工作。感谢他们给我机会完成我的第四本书并给予我支持。非常感谢大家。

本书的编写不是在孤立无援的状态下完成的。这里要感谢 Manuel Jordan 给我意见和建议，没有他的付出，本书的某些部分可能会有很大的不同。Manuel，谢谢你给我的意见、建议，以及你评审(和与我讨论)这本书的时间。

感谢我的家人和朋友，并再次感谢我的潜水伙伴们，我错过了所有的潜水和旅行。

还要感谢我的妻子 Djoke Deinum 以及女儿 Geeske 和 Sietske，感谢她们无穷的支持、爱和奉献。虽然我度过了很多漫长的夜晚写书，并牺牲了许多周末和假期来完成这本书，但如果没有她们的支持，我可能早就放弃了努力。

前　言

欢迎阅读《Spring Boot 2 攻略》。本书将专注于使用 Spring Boot 2.1 及其所支持的各种项目(如 Spring Security、Spring AMQP 等)进行软件开发。

本书读者对象

本书是为那些想要简化应用程序开发和快速学习编写应用程序的开发人员准备的。引入 Spring Boot 将简化应用程序配置，使用 Spring Boot 的全部功能还可以简化应用程序的部署和管理。

本书假定读者熟悉 Java、Spring 和某种 IDE。本书并没有解释 Spring 或相关项目的所有内部的、深入的工作原理。对于这些内容，请参阅 *Spring 5 Recipes* 或 *Pro Spring MVC*。

本书结构安排

第 1 章"介绍 Spring Boot"，简要介绍 Spring Boot 的特性以及如何创建 Spring Boot 项目。

第 2 章"Spring Boot 基础特性"，介绍如何定义和配置 bean 以及如何使用 Spring Boot 注入依赖项的基本场景。

第 3 章"Spring MVC 基础特性"，介绍如何使用 Spring MVC 开发基于 Web 的应用。

第 4 章"Spring MVC 异步特性"，介绍如何使用 Spring MVC 开发异步的 Web 应用。

第 5 章"Spring WebFlux 特性"，介绍如何使用 Spring WebFlux 开发反应式 Web 应用。

第 6 章"Spring Security 介绍"，简单说明如何使用 Spring Security 为 Spring Boot 应用程序提供安全保护。

第 7 章"数据访问",说明如何访问数据库、MongoDB 等数据存储。

第 8 章"Java 企业服务",介绍在 Spring Boot 中如何使用 JMX、Mail 和任务调度等企业级服务。

第 9 章"消息传递",介绍如何在 Spring Boot 中使用 JMS、RabbitMQ 实现传递消息。

第 10 章"Spring Boot Actuator",说明如何通过 Spring Boot Actuator 使用产品提供的各种特性,例如,通过管理端点来监视应用程序的健康状态和性能指标。

第 11 章"打包",介绍如何将 Spring Boot 应用程序构建为可执行文件或打包进 Docker 容器,以便打包和部署应用程序。

本书约定

有时候,当本书希望你将注意力集中到代码示例中的某个部分时,相应的代码会用粗体字显示。请注意粗体字部分并不一定表示这些代码修改了前面示例中的代码。

当代码行超过本书页面的宽度时,本书将使用字符连接字(-)连接分行的代码。当你在尝试输入代码时,请注意不要使用任何空格,直接连接输入即可。

阅读本书的前提条件

因为 Java 编程语言是独立于平台的,所以你可以自由选择任何受支持的操作系统。但是,本书中的一些示例使用了特定于平台的路径。在输入示例的代码之前,根据需要将它们转换为所选操作系统的格式。

要充分利用本书,请安装 JDK 11[1] 或更高版本。应该安装一个 Java IDE,以便于开发。对于本书,大多数示例代码都是基于 Maven[2] 的,并且大多数 IDE 都内置了对 Maven 管理类路径的支持。这些示例都使用了 Maven Wrapper[3],因此你不必安装 Maven 就可以从命令行构建示例。

这些示例有时需要额外的库,比如 PostreSQL、ActiveMQ 等,为此,本书使用了 Docker[4]。当然,你可以在机器上安装库,而不是使用 Docker,但是为了便于使用(并且不污染你的系统),最好使用 Docker。

---

1 https://adoptopenjdk.net/.
2 https://maven.apache.org/.
3 https://github.com/takari/maven-wrapper.
4 https://www.docker.com.

下载代码

本书的源代码可以通过 www.apress.com/9781484239629 下载，也可扫封底二维码获取源代码，源代码是按章节组织的，每个章节包括一个较独立的例子。

目 录

第 1 章 介绍 Spring Boot ·············1
  1.1 使用 Maven 创建 Spring Boot 应用程序 ···············2
    1.1.1 问题 ·······················2
    1.1.2 解决方案 ···············2
    1.1.3 工作原理 ···············2
  1.2 使用 Gradle 创建 Spring Boot 应用程序 ···············5
    1.2.1 问题 ·······················5
    1.2.2 解决方案 ···············5
    1.2.3 工作原理 ···············6
  1.3 使用 Spring Initializr 创建 Spring Boot 应用程序 ·······8
    1.3.1 问题 ·······················8
    1.3.2 解决方案 ···············9
    1.3.3 工作原理 ···············9
  1.4 小结 ·······························12

第 2 章 Spring Boot 基础特性 ·······13
  2.1 配置 bean ·······················13
    2.1.1 问题 ·····················13
    2.1.2 解决方案 ·············13
    2.1.3 工作原理 ·············13
  2.2 属性外置 ·······················19
    2.2.1 问题 ·····················19
    2.2.2 解决方案 ·············19
    2.2.3 工作原理 ·············19
  2.3 测试 ·······························23

    2.3.1 问题 ·····················23
    2.3.2 解决方案 ·············23
    2.3.3 工作原理 ·············23
  2.4 配置日志 ·······················28
    2.4.1 问题 ·····················28
    2.4.2 解决方案 ·············28
    2.4.3 工作原理 ·············28
  2.5 重用现有配置 ···············29
    2.5.1 问题 ·····················29
    2.5.2 解决方案 ·············30
    2.5.3 工作原理 ·············30

第 3 章 Spring MVC 基础特性 ······31
  3.1 开始使用 Spring MVC ·······31
    3.1.1 问题 ·····················31
    3.1.2 解决方案 ·············31
    3.1.3 工作原理 ·············31
  3.2 使用 Spring MVC 公开 REST 资源 ·························35
    3.2.1 问题 ·····················35
    3.2.2 解决方案 ·············35
    3.2.3 工作原理 ·············36
  3.3 在 Spring Boot 中使用 Thymeleaf 模板 ···············45
    3.3.1 问题 ·····················45
    3.3.2 解决方案 ·············45
    3.3.3 工作原理 ·············45
  3.4 处理异常 ·······················51

3.4.1　问题······51
　　　3.4.2　解决方案······51
　　　3.4.3　工作原理······51
　3.5　应用程序国际化······56
　　　3.5.1　问题······56
　　　3.5.2　解决方案······56
　　　3.5.3　工作原理······56
　3.6　解析用户区域设置······59
　　　3.6.1　问题······59
　　　3.6.2　解决方案······59
　　　3.6.3　工作原理······59
　3.7　选择和配置内嵌的
　　　服务器······63
　　　3.7.1　问题······63
　　　3.7.2　解决方案······63
　　　3.7.3　工作原理······63
　3.8　为 Servlet 容器配置 SSL······68
　　　3.8.1　问题······68
　　　3.8.2　解决方案······68
　　　3.8.3　工作原理······68

第 4 章　Spring MVC 异步特性······73

　4.1　使用控制器和 TaskExecutor
　　　处理异步请求······74
　　　4.1.1　问题······74
　　　4.1.2　解决方案······74
　　　4.1.3　工作原理······74
　4.2　响应回写函数······78
　　　4.2.1　问题······78
　　　4.2.2　解决方案······78
　　　4.2.3　工作原理······78
　4.3　WebSocket······86
　　　4.3.1　问题······86
　　　4.3.2　解决方案······86
　　　4.3.3　工作原理······86
　4.4　在 WebSocket 上使用
　　　STOMP······96
　　　4.4.1　问题······96

　　　4.4.2　解决方案······96
　　　4.4.3　工作原理······96

第 5 章　Spring WebFlux 特性······105

　5.1　使用 Spring WebFlux 开发
　　　反应式应用······105
　　　5.1.1　问题······105
　　　5.1.2　解决方案······105
　　　5.1.3　工作原理······107
　5.2　发布和使用反应式 Rest
　　　服务······111
　　　5.2.1　问题······111
　　　5.2.2　解决方案······111
　　　5.2.3　工作原理······111
　5.3　使用 Thymeleaf 作为模板
　　　引擎······120
　　　5.3.1　问题······120
　　　5.3.2　解决方案······120
　　　5.3.3　工作原理······120
　5.4　WebFlux 和 WebSocket······125
　　　5.4.1　问题······125
　　　5.4.2　解决方案······125
　　　5.4.3　工作原理······125

第 6 章　Spring Security 介绍······135

　6.1　在 Spring Boot 应用程序中
　　　启用安全特性······135
　　　6.1.1　问题······135
　　　6.1.2　解决方案······135
　　　6.1.3　工作原理······135
　6.2　登录 Web 应用······142
　　　6.2.1　问题······142
　　　6.2.2　解决方案······142
　　　6.2.3　工作原理······142
　6.3　用户认证······150
　　　6.3.1　问题······150
　　　6.3.2　解决方案······151
　　　6.3.3　工作原理······151

6.4	制定访问控制决策 ················ 156	
	6.4.1 问题 ····························· 156	
	6.4.2 解决方案 ····················· 156	
	6.4.3 工作原理 ····················· 156	
6.5	向 WebFlux 应用程序添加安全特性 ································ 160	
	6.5.1 问题 ····························· 160	
	6.5.2 解决方案 ····················· 161	
	6.5.3 工作原理 ····················· 161	
6.6	小结 ·· 166	

第 7 章 数据访问 ································ 167

- 7.1 配置数据源 ································ 167
  - 7.1.1 问题 ····························· 167
  - 7.1.2 解决方案 ····················· 167
  - 7.1.3 工作原理 ····················· 167
- 7.2 使用 JdbcTemplate ···················· 176
  - 7.2.1 问题 ····························· 176
  - 7.2.2 解决方案 ····················· 176
  - 7.2.3 工作原理 ····················· 176
- 7.3 使用 JPA ···································· 184
  - 7.3.1 问题 ····························· 184
  - 7.3.2 解决方案 ····················· 184
  - 7.3.3 工作原理 ····················· 184
- 7.4 直接使用 Hibernate ·················· 192
  - 7.4.1 问题 ····························· 192
  - 7.4.2 解决方案 ····················· 192
  - 7.4.3 工作原理 ····················· 192
- 7.5 Spring Data MongoDB ·············· 195
  - 7.5.1 问题 ····························· 195
  - 7.5.2 解决方案 ····················· 195
  - 7.5.3 工作原理 ····················· 195

第 8 章 Java 企业服务 ······················ 209

- 8.1 Spring 异步处理机制 ················ 209
  - 8.1.1 问题 ····························· 209
  - 8.1.2 解决方案 ····················· 209
  - 8.1.3 工作原理 ····················· 209

- 8.2 Spring 任务调度 ························ 213
  - 8.2.1 问题 ····························· 213
  - 8.2.2 解决方案 ····················· 213
  - 8.2.3 工作原理 ····················· 213
- 8.3 发送 E-mail ······························· 215
  - 8.3.1 问题 ····························· 215
  - 8.3.2 解决方案 ····················· 215
  - 8.3.3 工作原理 ····················· 215
- 8.4 注册 JMX MBean ····················· 220
  - 8.4.1 问题 ····························· 220
  - 8.4.2 解决方案 ····················· 220
  - 8.4.3 工作原理 ····················· 220

第 9 章 消息传递 ································ 225

- 9.1 配置 JMS ·································· 225
  - 9.1.1 问题 ····························· 225
  - 9.1.2 解决方案 ····················· 225
  - 9.1.3 工作原理 ····················· 225
- 9.2 使用 JMS 发送消息 ·················· 231
  - 9.2.1 问题 ····························· 231
  - 9.2.2 解决方案 ····················· 231
  - 9.2.3 工作原理 ····················· 231
- 9.3 使用 JMS 接收消息 ·················· 238
  - 9.3.1 问题 ····························· 238
  - 9.3.2 解决方案 ····················· 238
  - 9.3.3 工作原理 ····················· 238
- 9.4 配置 RabbitMQ ························· 242
  - 9.4.1 问题 ····························· 242
  - 9.4.2 解决方案 ····················· 242
  - 9.4.3 工作原理 ····················· 243
- 9.5 使用 RabbitMQ 发送消息 ································ 243
  - 9.5.1 问题 ····························· 243
  - 9.5.2 解决方案 ····················· 244
  - 9.5.3 工作原理 ····················· 244
- 9.6 使用 RabbitMQ 接收消息 ································ 251
  - 9.6.1 问题 ····························· 251

9.6.2 解决方案······251
9.6.3 工作原理······251

第 10 章　Spring Boot Actuator······255
10.1 启用和配置 Spring Boot Actuator······255
10.1.1 问题······255
10.1.2 解决方案······255
10.1.3 工作原理······255
10.2 创建自定义的健康状况检查和性能指标······261
10.2.1 问题······261
10.2.2 解决方案······262
10.2.3 工作原理······262
10.3 导出性能指标······264
10.3.1 问题······264
10.3.2 解决方案······264
10.3.3 工作原理······265

第 11 章　打包······267
11.1 创建可执行文件······267
11.1.1 问题······267
11.1.2 解决方案······267
11.1.3 工作原理······267
11.2 为部署创建 WAR 文件···270
11.2.1 问题······270
11.2.2 解决方案······270
11.2.3 工作原理······270
11.3 通过 Thin Launcher 减少归档文件大小······273
11.3.1 问题······273
11.3.2 解决方案······273
11.3.3 工作原理······273
11.4 使用 Docker······274
11.4.1 问题······275
11.4.2 解决方案······275
11.4.3 工作原理······275

第 1 章

介绍 Spring Boot

本章将简要介绍 Spring Boot。Spring Boot 的核心是 Spring 框架；Spring Boot 扩展了这个框架，使自动配置成为可能。

> Spring Boot 使得创建独立的、生产级的、基于 Spring 的应用程序变得简单，你可以"直接运行"这些应用程序。我们从一个独到的视角学习 Spring 平台和第三方库，这样你就能以最小的麻烦开始了。大多数 Spring Boot 应用程序只需要很少的 Spring 配置。
>
> ——《Spring Boot 参考指南》

Spring Boot 为众多基础设施(如 JMS、JDBC、JPA、RabbitMQ)提供了自动配置。Spring Boot 还为不同的框架(如 Spring Integration、Spring Batch、Spring Security)提供自动配置。当检测到这些框架或功能时，Spring Boot 将使用固定但合理的默认值来配置它们。

本书源代码使用 Maven 进行构建。Maven 将负责获取必要的依赖项、编译代码和创建工件(通常是 JAR 文件)。

■ 提示：

要构建某个应用程序时，进入相应的 Recipe 目录(如 ch2/recipe_2_1_i/)并执行 mvnw 命令来编译源代码。源代码编译完成后，将为应用程序的可执行文件创建 target 子目录。然后，可以从命令行运行应用程序 JAR 包(例如，运行命令 java-jar target/Recipe_2_1_i.jar)。

1.1 使用 Maven 创建 Spring Boot 应用程序

1.1.1 问题

你希望开始使用 Spring Boot 和 Maven 开发应用程序。

1.1.2 解决方案

创建一个 Maven 构建文件，即 pom.xml，并添加所需的依赖项。要启动应用程序，创建一个 Java 类，其中包含启动应用程序的 main 方法。

1.1.3 工作原理

假设你要创建一个简单的应用程序，它启动一个 SpringApplication(Spring Boot 应用程序的主要入口点)，从 ApplicationContext 获取所有 bean 的名称，并将这些名称输出到控制台。

1. 创建 pom.xml

在开始编码之前，需要创建 pom.xml 文件，Maven 使用该文件来确定需要做什么。使用 Spring Boot 的最简单方法是将 spring-boot-starter-parent 作为应用程序的 parent 节点。

```xml
<parent>
    <groupId>org.springframework.boot</groupId>
    <artifactId>spring-boot-starter-parent</artifactId>
    <version>2.1.0.RELEASE</version>
    <relativePath />
</parent>
```

接下来，需要添加一些 Spring 依赖项以开始使用 Spring。为此，将 spring-boot-starter 作为依赖项添加到 pom.xml 中。

```xml
<dependencies>
    <dependency>
        <groupId>org.springframework.boot</groupId>
        <artifactId>spring-boot-starter</artifactId>
    </dependency>
</dependencies>
```

请注意，不需要任何版本或其他信息，所有这些都是被管理的，因为将 spring-boot-starter-parent 作为应用程序的父节点。spring-boot-starter 将引入启动一个非常基本的 Spring Boot 应用程序所需的所有核心依赖项，比如 Spring 框架、用于记录日志的 Logback 以及 Spring Boot 本身。

完整的 pom.xml 文件现在应该如下所示：

```xml
<?xml version="1.0" encoding="UTF-8"?>
<project xmlns="http://maven.apache.org/POM/4.0.0"
         xmlns:xsi="http://www.w3.org/2001/XMLSchema-instance"
         xsi:schemaLocation="http://maven.apache.org/POM/4.0.0
         http://maven.apache.org/xsd/maven-4.0.0.xsd">

    <modelVersion>4.0.0</modelVersion>
    <groupId>com.apress.springbootrecipes</groupId>
    <artifactId>chapter_1_1</artifactId>
    <version>2.0.0</version>

    <parent>
        <groupId>org.springframework.boot</groupId>
        <artifactId>spring-boot-starter-parent</artifactId>
        <version>2.1.0.RELEASE</version>
    </parent>

    <dependencies>
      <dependency>
        <groupId>org.springframework.boot</groupId>
        <artifactId>spring-boot-starter</artifactId>
      </dependency>
    </dependencies>
</project>
```

2. 创建应用程序类

让我们创建一个拥有 main 方法的 DemoApplication 类。main 方法使用 DemoApplication.class 和 main 方法中的参数来调用 SpringApplication.run。run 方法返回一个 ApplicationContext 对象，用于从 ApplicationContext 中检索 bean 名称。将对象名称进行排序，然后打印到控制台。

生成的类如下所示：

```java
package com.apress.springbootrecipes.demo;

import org.springframework.boot.SpringApplication;
```

```java
import org.springframework.boot.autoconfigure.EnableAutoConfiguration;
import org.springframework.context.annotation.ComponentScan;
import org.springframework.context.annotation.Configuration;

import java.util.Arrays;

@Configuration
@EnableAutoConfiguration
@ComponentScan
public class DemoApplication {

    public static void main(String[] args) {
        var ctx = SpringApplication.run(DemoApplication.class, args);

        System.out.println("# Beans:"+ctx.getBeanDefinitionCount());

        var names = ctx.getBeanDefinitionNames();
        Arrays.sort(names);
        Arrays.asList(names).forEach(System.out::println);
    }
}
```

这个类是一个带有 main 方法的普通 Java 类，可以从 IDE 运行这个类。当应用程序运行时，它的输出将与图 1-1 类似。

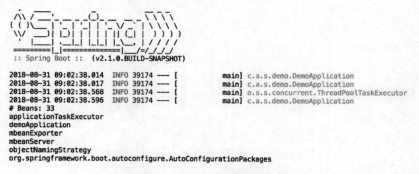

图 1-1　运行应用程序的输出

代码和注解发生了什么？@Configuration 注解使这个类成为 Spring Java 配置类。通常，在创建应用程序时，你还需要使用其他组件，为此，请添加 @ComponentScan 注解。最后，要让 Spring Boot 执行其自动配置，请添加 @Enable-AutoConfiguration 注解。

3. 简化应用程序类

如果查看类的定义，可以看到有三条注解。

```
@Configuration
@EnableAutoConfiguration
@ComponentScan
public class DemoApplication { ... }
```

在编写基于 Spring Boot 的应用程序时，大多数应用程序都需要所有这些注解。可使用@SpringBootApplication 注解来简化这些代码。随后，类的头部变为：

```
@SpringBootApplication
public class DemoApplication { ... }
```

@SpringBootApplication 注解是一个所谓的组合注解，包含了前面需要的几个注解。

```
@Target({ElementType.TYPE})
@Retention(RetentionPolicy.RUNTIME)
@Documented
@Inherited
@SpringBootConfiguration
@EnableAutoConfiguration
@ComponentScan
public @interface SpringBootApplication { ... }
```

@SpringBootApplication 和前面提到的多个注解之间有一个区别。这里使用@SpringBootConfiguration注解而不是@Configuration注解。@SpringBootConfiguration是一个专门的@Configuration 注解。它表明这是一个基于 Spring Boot 的应用程序。在应用程序中使用@SpringBootConfiguration 时，只能有一个类用此注解进行注解！

1.2 使用 Gradle 创建 Spring Boot 应用程序

1.2.1 问题

你希望开始使用 Spring Boot 和 Gradle 开发应用程序。

1.2.2 解决方案

创建 Gradle 构建文件 build.gradle，并添加所需的依赖项。为了启动应用程序，创建一个 Java 类，其中包含一个 main 方法来引导应用程序。

1.2.3 工作原理

假设你要创建一个简单的应用程序，它启动一个 SpringApplication，从 ApplicationContext 中获取所有 bean 的名称，并将它们输出到控制台。

1. 创建 build.gradle

首先，需要创建 build.gradle 文件并使用 Gradle 所需的两个插件来正确管理 Spring Boot 的依赖项。Spring Boot 需要一个特殊的 Gradle 插件(Spring Boot Gradle 插件)和一个插件(依赖管理插件)来扩展 Gradle 默认的依赖管理功能。要启用和配置这些插件，需要在 build.gradle 中创建一个 buildscript 任务。

```
buildscript {
  ext {
       springBootVersion = '2.1.0.RELEASE'
  }
  repositories {
     mavenCentral()
  }
  dependencies {
      classpath("org.springframework.boot:spring-boot-gradle-plugin:$ {springBootVersion}")
  }
}
```

现在此任务将正确配置要使用的 Spring Boot 插件。接下来，你需要指定要使用的插件；因为这是一个基于 Java 的项目，你至少需要 Java 插件；因为本书是关于 Spring Boot 的，你还需要 org.springframework.boot 插件。最后，你需要包含 io.spring.dependency-management 插件，以便让多个 Spring Boot Starter 管理依赖项。

```
apply plugin: 'java'
apply plugin: 'org.springframework.boot'
apply plugin: 'io.spring.dependency-management'
```

最后，将需要添加所需的依赖项，添加 spring-boot-starter 依赖项。

```
dependencies {
    compile 'org.springframework.boot:spring-boot-starter'
}
```

请注意，依赖项上没有特定的版本号。不需要指定版本，版本的管理会自动进行，这是由于使用了 io.spring.dependency-management 插件。与 Maven 一样，该插件使得依赖管理更方便。

完整的 build.gradle 文件现在应该如下所示:

```
buildscript {
    ext {
        springBootVersion = '2.1.0.RELEASE'
    }
    repositories {
        mavenCentral()
    }
    dependencies {
        classpath("org.springframework.boot:spring-boot-gradle-
        plugin:$ {springBootVersion}")
    }
}

apply plugin: 'java'
apply plugin: 'org.springframework.boot'
apply plugin: 'io.spring.dependency-management'

dependencies {
    compile 'org.springframework.boot:spring-boot-starter'
}
repositories {
    mavenCentral()
}
```

2. 创建应用程序类

下面创建一个拥有 main 方法的 DemoApplication 类。main 方法使用 DemoApplication.class 和 main 方法中的参数调用 SpringApplication.run。run 方法返回一个 ApplicationContext 对象,用于从 ApplicationContext 中检索 bean 的名称。将对象名称进行排序,然后打印到控制台。

生成的类如下所示:

```
package com.apress.springbootrecipes.demo;

import org.springframework.boot.SpringApplication;
import org.springframework.boot.autoconfigure.SpringBootApplication;
```

```java
import java.util.Arrays;

@SpringBootApplication
public class DemoApplication {

    public static void main(String[] args) {
        var ctx = SpringApplication.run(DemoApplication.class, args);

        System.out.println("# Beans:"+ctx.getBeanDefinitionCount());

        var names = ctx.getBeanDefinitionNames();
        Arrays.sort(names);
        Arrays.asList(names).forEach(System.out::println);
    }
}
```

这个类是一个带有 main 方法的普通 Java 类。你可以从 IDE 运行这个类。当应用程序运行时，它的输出将与图 1-2 类似。

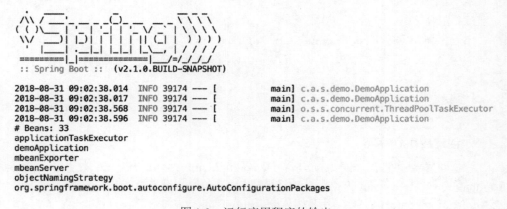

图 1-2 运行应用程序的输出

1.3 使用 Spring Initializr 创建 Spring Boot 应用程序

1.3.1 问题

你希望开始使用 Spring Initializr 开发 Spring Boot 应用程序。

1.3.2 解决方案

在浏览器中打开 http://start.spring.io 页面，选择 Spring Boot 版本和你认为需要的依赖项，然后下载该项目。

1.3.3 工作原理

首先在浏览器中打开 http://start.spring.io 页面，进入 Spring Initializr 界面，如图 1-3 所示。

现在选择要生成的内容(Maven 或 Gradle)。选择要使用的 Spring Boot 版本，请选择最新版本。接下来在 Group 字段中输入 com.apress.springbootrecipes.demo，在 Artifact 字段中保留默认的 demo 值，如图 1-4 所示。

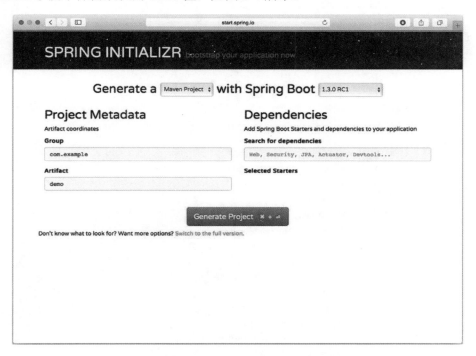

图 1-3　Spring Initializr 界面

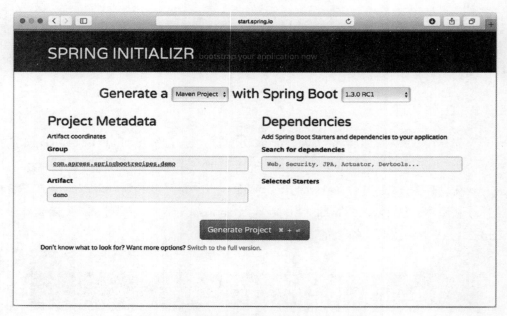

图 1-4　设置好参数值的 Spring Initializr 界面

最后，单击 Generate Project 按钮，这将触发下载 demo.zip。解压此 zip 文件并将项目导入 IDE。导入后，你应该看到一个类似于图 1-5 所示的结构。

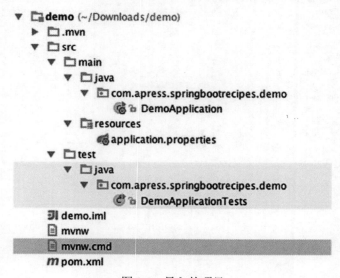

图 1-5　导入的项目

打开 pom.xml 文件，并将其与 1.1 节中的 pom.xml 文件(或 1.2 节中的 build.gradle

文件)进行比较。可以看到它们非常相似，然而，有两个不同之处需要注意。首先，pom.xml 文件有一个附加的依赖项，即 spring-boot-starter-test。这将引入所需的测试依赖项，如 Spring Test、Mockito、JUnit 4 和 AssertJ 等。有了这个单一的依赖关系，你就可以开始测试了。

第二个区别是现在有一个配置了 spring-boot-maven-plugin 的构建块。

```xml
<build>
  <plugins>
    <plugin>
      <groupId>org.springframework.boot</groupId>
      <artifactId>spring-boot-maven-plugin</artifactId>
    </plugin>
  </plugins>
</bild>
```

这个插件负责创建一个胖 JAR。它获取原始 JAR 文件，并将其与内部的所有依赖项重新打包。这样，你就可以将 JAR 文件移交给操作团队。操作团队需要执行命令 java -jar <your-application>.jar 来启动应用程序。不需要将其部署到 Servlet 容器或 JEE 容器中。

1. 实现一个简单的应用程序

打开 DemoApplication 并修改内容，以便从 ApplicationContext 获取 bean 的数量和 bean 的名称清单。

```java
package com.apress.springbootrecipes.demo;

import org.springframework.boot.SpringApplication;
import org.springframework.boot.autoconfigure.SpringBootApplication;

import java.util.Arrays;

@SpringBootApplication
public class DemoApplication {

    public static void main(String[] args) {
        var ctx = SpringApplication.run(DemoApplication.class, args);

        System.out.println("# Beans:"+ctx.getBeanDefinitionCount());

        var names = ctx.getBeanDefinitionNames();
```

```
        Arrays.sort(names);
        Arrays.asList(names).forEach(System.out::println) }
    }
}
```

2. 构建 JAR

在使用 Spring Initializr 时，所有项目都带有 Maven 打包器(或使用 Gradle 时的 Gradle 打包器)，以便于构建应用程序。要使用打包脚本，请打开命令行；导航到项目所在的目录；最后，执行 ./mvnw package 或/gradlew build。这应该在 target(或 build/libs)目录中创建可执行的 JAR 包。

现在已经构建了 JAR，那么让我们执行它，看看会发生什么。输入命令java-jar target/demo-0.0.1-SNAPSHOT.jar(或者 java -jar build/libs/demo-0.0.1-SNAPSHOT.jar)，查看应用程序的启动并从上下文中列出 bean 的名称(参见图 1-1 和图 1-2)。

1.4 小结

本章介绍了如何使用 Spring Boot 引导开发。我们研究了如何开始使用 Maven 和 Gradle，最后研究了如何开始使用 Spring Initializr。

在第 2 章中将介绍 Spring Boot 应用程序的基础配置，如何定义 bean，如何使用属性文件以及如何覆盖属性。

第 2 章

Spring Boot 基础特性

在本章中，我们将介绍 Spring Boot 的基础特性。

■ 注意：
作为开始，使用 Spring Initializr 创建一个项目。创建项目时不需要添加其他的依赖关系，只是一个单纯的 Spring Boot 项目。

2.1 配置 bean

2.1.1 问题

你希望 Spring Boot 将类作为 bean 使用。

2.1.2 解决方案

根据需要，可以利用 @ComponentScan 自动地检测类并创建类的实例，该注解与 @Autowired 和 @Value 一起使用以获取依赖项或注入的属性；或者可以使用注解 @Bean，这种方法可以对正在创建的 bean 的构造过程获取更多的控制。

2.1.3 工作原理

1.1 小节中说明了 @SpringBootApplication 包括 @ComponentScan 和 @Configuration。这意味着任何使用 @Component 注解的类都将由 Spring Boot 自动检测和实例化，还允许使用 @Bean 来注解方法，从而将方法声明为 bean。

1. 使用 @Component 注解

首先，创建一个用于启动应用程序的类，这里创建一个用 @SpringBootApplication 注解的 HelloWorldApplication 类。

```
@SpringBootApplication
```

```
public class HelloWorldApplication {

  public static void main(String[] args) {
      SpringApplication.run(HelloWorldApplication.class, args);
  }
}
```

■ 提示：
需要将@SpringBootApplication注解的类放在顶层包中，这样它将自动检测在顶层包和所有子包中定义的所有带注解的组件、配置类等[1]。

这个类将启动应用程序，检测所有使用@Component注解的类，并检测类路径上包含了哪些库。当运行这个 HelloWorldApplication 时，它不会做很多事情，因为没有什么可以检测或运行的。让我们创建一个简单的类，它将由 Spring Boot 自动检测。

```
@Component
public class HelloWorld {

  @PostConstruct
  public void sayHello() {
    System.out.println("Hello World, from Spring Boot 2!");
  }
}
```

Spring Boot 将检测到这个类并从中创建一个 bean 实例。在创建好实例和注入所有的依赖项之后，Spring Boot 会调用由@PostConstruct注解的方法。简单地说，在应用程序启动时，sayHello 方法将会运行，控制台将输出行"Hello World, from Spring Boot 2!"。

■ 重要：
当需要对默认的组件扫描过程未覆盖的包进行扫描时，需要在@SpringBootApplication注解类中添加使用@ComponentScan注解的包名称，以便对这些包进行扫描。添加完成后，除了因@SpringBootApplication而进行的默认扫描之外，这些包也会被扫描。

---

1 https://docs.spring.io/spring-boot/docs/current/reference/htmlsingle/#using-boot-locating-the-main-class。

2. 使用由@Bean 注解的方法

除了通过自动检测组件的方式以外，也可以使用工厂方法来创建 bean。如果想要或需要更多地控制 bean 的构造过程，这种方式是很有用的。工厂方法是使用 @Bean 注解的方法[2]，它将用于在应用程序上下文中注册 bean。bean 的名称与方法的名称相同。方法可以有参数，应用程序上下文将为其他 bean 解析这些参数。

让我们创建一个可以对整数进行一些基本计算的应用程序。首先，编写 Calculator 类，它将在构造函数中获得一组 Operation bean。Operation 是一个接口，多个不同的实现将进行实际的计算。

```
package com.apress.springbootrecipes.calculator;

import java.util.Collection;

public class Calculator {

    private final Collection<Operation> operations;

    public Calculator(Collection<Operation> operations) {
        this.operations = operations;
    }

    public void calculate(int lhs, int rhs, char op) {
        for (var operation : operations) {
            if (operation.handles(op)) {
                var result = operation.apply(lhs, rhs);
                System.out.printf("%d %s %d = %s%n",lhs,op,rhs,result);
                return;
            }
        }
        throw new IllegalArgumentException("Unknown operation " + op);
    }
}
```

在 calculate 方法中，使用 Operation.handles 方法检测正确的 Operation；找到正确的 Operation 后，调用 Operation.apply 方法进行实际计算。如果传入计算器的操作无法处理，则会引发异常。

Operation 接口是一个简单的接口，包含前面提到的两个方法。

---

[2] https://docs.spring.io/spring/docs/current/spring-framework-reference/core.html#Beans-java-Bean-annotation.

```
package com.apress.springbootrecipes.calculator;

public interface Operation {

    int apply(int lhs, int rhs);
    boolean handles(char op);
}
```

现在添加两个操作：一个用于值相加，另一个用于值相乘。

```
package com.apress.springbootrecipes.calculator.operation;

import com.apress.springbootrecipes.calculator.Operation;
import org.springframework.stereotype.Component;

@Component
class Addition implements Operation {

    @Override
    public int apply(int lhs, int rhs) {
        return lhs + rhs;
    }
    @Override
    public boolean handles(char op) {
        return '+' == op;
    }
}
package com.apress.springbootrecipes.calculator.operation;

import com.apress.springbootrecipes.calculator.Operation;
import org.springframework.stereotype.Component;

@Component
class Multiplication implements Operation {

    @Override
    public int apply(int lhs, int rhs) {
        return lhs * rhs;
    }
```

```java
    @Override
    public boolean handles(char op) {
        return '*' == op;
    }
}
```

以上就是制作一个计算器需要的所有组件,它可以进行加法和乘法运算,并且可对其进行扩展。

最后,让我们编写一个应用程序来配置和使用 Calculator。要创建 Calculator 的实例,需要一个使用 @Bean 注解的方法。此方法可以添加到使用 @SpringBoot-Application 或常规的 @Configuration 注解的类中。

```java
package com.apress.springbootrecipes.calculator;

import org.springframework.boot.SpringApplication;
import org.springframework.boot.autoconfigure.SpringBootApplication;
import org.springframework.context.annotation.Bean;

import java.util.Collection;

@SpringBootApplication
public class CalculatorApplication {
    public static void main(String[] args) {

        var ctx = SpringApplication.run(CalculatorApplication.class, args);

        var calculator = ctx.getBean(Calculator.class);
        calculator.calculate(137, 21, '+');
        calculator.calculate(137, 21, '*');
        calculator.calculate(137, 21, '-');
    }

    @Bean
    public Calculator calculator(Collection<Operation> operations) {
        return new Calculator(operations);
    }
}
```

calculator 工厂方法接收参数 List<Operation>，并使用这个参数来构造 Calculator 实例。当在@Bean 注解的方法中使用参数时，这些参数将被自动解析。当注入集合时，Spring 将自动检测所需 bean 的所有实例，并使用这个集合来调用 calculator 工厂方法。

在 main 方法中，代码获取 Calculator 类的实例并使用不同的数字和操作调用其 calculate 方法。前两次调用将正确地将一些输出打印到控制台，最后一次调用将抛出一个异常，因为没有合适的操作执行减法运算。

虽然为 Calculator 创建了一个工厂方法，但实际上并不需要。当使用 @Component 注解 Calculator 类时，Spring 会检测到该方法。类中仅有的构造函数将用于构造 Calculator 的实例。当有多个构造函数时，使用@Autowired 注解需要使用的构造函数。

这段代码另一个有欠缺的地方是，bean 是手动检索的，一般认为这是一种糟糕的编程方法，更合理的方式是使用依赖注入。Spring Boot 有一个接口 ApplicationRunner，它可以用于在应用程序启动后运行一些代码。当 Spring Boot 检测到 ApplicationRunner 类型的 bean 时，它将在应用程序启动后立即调用它的 run 方法。

```java
package com.apress.springbootrecipes.calculator;

import org.springframework.boot.ApplicationRunner;
import org.springframework.boot.SpringApplication;
import org.springframework.boot.autoconfigure.SpringBootApplication;
import org.springframework.context.annotation.Bean;

@SpringBootApplication
public class CalculatorApplication {

  public static void main(String[] args) {
     SpringApplication.run(CalculatorApplication.class, args);
  }

  @Bean
  public ApplicationRunner calculationRunner(Calculator calculator) {
    return args -> {
      calculator.calculate(137, 21, '+');
      calculator.calculate(137, 21, '*');
      calculator.calculate(137, 21, '-');
    };
  }
}
```

创建 Calculator 类的方法已被 ApplicationRunner 替代。该 ApplicationRunner 接收自动配置好的 Calculator 类并在该类上运行一些操作。当运行这个类时，输出应该和以前一样。主要的区别和优势在于不再需要从 ApplicationContext 手动获取 bean，因为 Spring 将负责获取正确的 bean。

2.2 属性外置

2.2.1 问题

你希望使用属性为不同的环境或运行场景配置应用程序。

2.2.2 解决方案

默认情况下，Spring Boot 支持从多个位置获取属性。它将默认加载名为 application.properties 的文件，并使用环境变量和 Java System 属性。从命令行运行应用程序时，它还将考虑命令行参数。根据应用程序的类型和可用的功能(例如安装了 JNDI)，Spring Boot 可以考虑从更多的位置加载属性[3]。对于我们的应用程序，Spring Boot 按照如下顺序考虑以下资源：

(1) 命令行参数
(2) 应用程序包之外的 application.properties 文件
(3) 应用程序包中的 application.properties 文件

除此之外，对于选项(2)和(3)，还可以基于当前激活的配置(active profile)来加载特定的配置文件。需要激活的配置文件可以通过 spring.profiles.active 属性传递。特定于配置的 application-{profile}.properties 文件优先于非特定于配置的文件。每一个配置文件都将被加载，可以使用这些文件覆盖属性，这使得属性优先级的清单比较长。

(1) 命令行参数
(2) 应用程序包之外的 application-{profile}.properties
(3) 应用程序包之外的 application.properties
(4) 应用程序包中的 application-{profile}.properties
(5) 应用程序包中的 application.properties

2.2.3 工作原理

在 2.1 节中创建的 Calculator 类是非常灵活的，但是，CalculatorApplication 类在进行计算时使用了硬编码的值。现在，如果需要进行其他的计算，就需要修改代码，重新编译，并运行新编译的代码。我们希望使用属性完成配置，以便在有需要时更改属性即可。

---

[3] https://docs.spring.io/spring-boot/docs/current/reference/html/boot-featuresexternal-config.html#boot-features-external-config.

首先修改应用程序，使其使用属性值而不是硬编码值。为此，更改 Application-Runner 类中用@Bean 注解的方法以接收 3 个额外的参数，这些参数将用@Value 注解。

```
package com.apress.springbootrecipes.calculator;

import org.springframework.beans.factory.annotation.Value;
import org.springframework.boot.ApplicationRunner;
import org.springframework.boot.SpringApplication;
import org.springframework.boot.autoconfigure.SpringBootApplication;
import org.springframework.context.annotation.Bean;

@SpringBootApplication
public class CalculatorApplication {

  public static void main(String[] args) {

    SpringApplication.run(CalculatorApplication.class, args);
  }

  @Bean
  public ApplicationRunner calculationRunner(Calculator calculator,
                    @Value("${lhs}") int lhs,
                    @Value("${rhs}") int rhs,
                    @Value("${op}") char op) {
    return args -> calculator.calculate(lhs, rhs, op);
  }
}
```

@Value 注解将指示 Spring 查找具体的属性并使用该属性的值。例如，如果我们使用@Value("${lhs}")，Spring 将尝试检测名为 lhs 的属性并使用该属性的值。还可以通过添加分号指定一个默认值。对于@Value("${lhs:12}")，如果找不到属性的值，则使用 12。如果没有指定默认值，Spring 将因为没有找到属性而引发 IllegalArgumentException 异常。如果现在启动应用程序，就会抛出异常，说明无法找到 lhs 属性。

在 src/main/resources 目录中添加 application.properties 文件，并将 lhs、rhs 和 op 三个属性的值保存在该文件中。

```
lhs=7
rhs=6
```

```
op=*
```

Spring Boot 将在启动时加载这个 application.properties 文件,此时已经可以访问属性。现在,运行应用程序时,它应该会再次生成与 application.properties 文件中给定的属性值相符合的输出。

尽管现在已经将属性外置到 application.properties 文件中,但这些属性仍然打包在应用程序内,这意味着为了执行其他计算仍然需要修改这些属性。每添加一种属性组合,都意味着需要重新编译一次代码。如果一个真正的生产系统采取这种方式,仅仅因为配置需要更改就重新编译代码,这是难以想象的。Spring Boot 有多种方式可以解决这个问题。

1. 使用外部 application.properties 文件进行覆盖

首先编译代码并使用命令 java-jar recipe_2_2--1.0.0.jar 启动应用程序,它将再次运行并加载前面提供的、打包在应用程序内的 application.properties 文件。现在,在与编译文件位置相同的目录下再添加一个 application.properties 文件,并将不同属性的值放入其中。

```
lhs=26
rhs=952
op=*
```

再次启动应用程序时,它将使用此 application.properties 文件中的属性值。

2. 使用配置文件覆盖属性

Spring Boot 可以使用激活的配置(active profile)加载额外的配置文件,这些配置文件可以完全替换或部分覆盖常规的配置。在 src/main/resources 目录中添加一个 application-add.properties 文件,它包含一个不同的 op 属性值。

```
op=+
```

现在编译代码(生成 JAR 文件)并在命令行通过命令 java -jar recipe_2_2-1.0.0.jar --spring.profiles.active=add 启动应用程序;应用程序将启动并加载 application.properties 文件和 application-add.properties 文件中的属性来配置应用程序。请注意这里执行的是加法而不是乘法,这表示 application-add.properties 文件优先于常规的 application.properties 文件。

> ■ 提示:
> 在使用未打包在应用程序内的 application.properties 文件和 application-{profile}.properties 文件时,上述规则也适用。

3. 使用命令行参数覆盖属性

最后一种方式是使用命令行参数来覆盖属性，上一节已经使用了命令行参数 --spring.profile.active=add 来指定激活的配置。也可以使用相同的方式来指定 lhs 和其他参数。使用命令 java -jar recipe_2_2-1.0.0.jar --lhs=12 --rhs=15 --op=+ 来运行应用程序，你将看到它根据通过命令行传递的参数进行计算。命令行中的参数始终会覆盖所有其他配置。

4. 从不同的配置文件加载属性

如果使用的是 application.properties 以外的其他文件，或者一些组件中内嵌的文件需要加载，则总是可以在 @SpringBootApplication 注解的类上使用额外的 @PropertySource 注解来加载指定的附加文件。

```
@PropertySource("classpath:your-external.properties")
@SpringBootApplication
public class MyApplication { ... }
```

@PropertySource 注解允许应用程序添加需要在启动期间加载的其他属性文件。除了使用@PropertySource 注解之外，还可以指示 Spring Boot 使用表 2-1 中的命令行参数加载其他属性文件。

表 2-1 配置参数

参数	说明
spring.config.name	要加载的文件名字符串，多个文件名用逗号分隔，默认值为 application
spring.config.location	资源位置(或文件名)字符串，多个资源位置(或文件名)之间用逗号分隔；应用程序将从这些资源位置(或文件)加载属性；默认值为 classpath:/、classpath:/config/、file:./、file:./config/
spring.config.additional-location	其他资源位置(或文件名)字符串，多个资源位置(或文件名)之间用逗号分隔；应用程序将从这些资源位置(或文件)加载属性；默认认为空

■ 提示：

当使用 spring.config.location 或 spring.config.additional-location 指向文件时，Spring Boot 将按照文件中存放的属性进行加载，并且不会加载特定于配置(profile-specific)的文件。当使用这两个属性指向目录时，Spring Boot 将加载特定于配置的文件。

在命令行通过指定参数--spring.config.name=your-external 来加载 your-external.properties 文件是可以的，但这会破坏加载 application.properties 文件的过程，更好的方式是指定--spring.config.name=application,your-external。现在将搜索所有的目录以查找 application.properties 和 your-external.properties，并加载特定于配置的文件，根据属性来源的优先级确定使用哪个属性值。

2.3 测试

2.3.1 问题

你希望为组件或 Spring Boot 应用程序的部分内容编写测试程序。

2.3.2 解决方案

Spring Boot 扩展了 Spring Test 框架的功能范围。它新增的特性支持对 bean 进行模拟和监视，并为 Web 测试提供了自动配置功能。同时，它也引入了简单的方法来测试应用程序的片段，只需要启动待测试的程序片段(例如，通过使用@WebMvcTest 或@JdbcTest)即能完成测试。

2.3.3 工作原理

Spring Boot 还将自动配置功能扩展到了测试框架的某些部分。它还与 Mockito 框架[4]集成，以便在 bean 上轻松地进行模拟(或监视)。此外，它还使用 Spring MockMvc 测试框架或基于 WebDriver 的测试为基于 Web 的测试提供自动配置功能。

1. 编写单元测试

首先，为计算器的一个组件编写简单的单元测试，即 MultiplicationTest，它将测试 Multiplication 类。

```java
public class MultiplicationTest {

  private final Multiplication addition = new Multiplication();

  @Test
  public void shouldMatchOperation() {
    assertThat(addition.handles('*')).isTrue();
    assertThat(addition.handles('/')).isFalse();
  }
```

---

[4] https://site.mockito.org.

```
@Test
public void shouldCorrectlyApplyFormula() {
    assertThat(addition.apply(2, 2)).isEqualTo(4);
    assertThat(addition.apply(12, 10)).isEqualTo(120);
  }
}
```

这是一个基本的单元测试。实例化Multiplication类,并调用它的方法,然后验证结果。第一段测试代码将测试它是否支持*操作符,第二段测试代码将测试它的乘法逻辑是否正确。要使一个方法成为测试方法(对于JUnit4框架而言),必须使用@Test对其进行注解。

2. 在单元测试中模拟依赖项

类有时需要依赖于其他内容,但你希望(在编写单元测试时)仅针对一个组件进行测试。Spring Boot会自动引入Mockito框架,该框架对于模拟类并记录其行为非常方便。为Calculator编写测试需要额外的组件,因为它将实际的计算委托给可用的Operation类。为测试Calculator的正确行为,我们需要创建一个Operation接口的模拟实例并将其注入Calculator。

```
public class CalculatorTest {

  private Calculator calculator;
  private Operation mockOperation;

  @Before
  public void setup() {
      mockOperation = Mockito.mock(Operation.class);
      calculator = new Calculator(Collections.
                  singletonList(mockOperation));
  }

  @Test(expected = IllegalArgumentException.class)
  public void throwExceptionWhenNoSuitableOperationFound() {

    when(mockOperation.handles(anyChar())).thenReturn(false);
    calculator.calculate(2, 2, '*');
  }

  @Test
```

```
public void shouldCallApplyMethodWhenSuitableOperationFound() {

    when(mockOperation.handles(anyChar())).thenReturn(true);
    when(mockOperation.apply(2, 2)).thenReturn(4);
    calculator.calculate(2, 2, '*');

    verify(mockOperation, times(1)).apply(2,2);
  }
}
```

在由@Before 注解的类中，我们通过调用 Mockito.mock 方法模拟 Operation 接口，并使用模拟的操作构造 Calculator 类的实例。该模拟实例在测试方法中执行特定操作。在第一个测试方法中，我们希望测试没有找到合适操作的情况，因此当调用 handles 方法时，我们指示模拟实例返回 false。测试预期将出现异常，如果抛出异常，则测试成功。第二个测试是检查模拟实例是否遵循了正确的流程，测试其行为是否正确。我们指示模拟实例为 handles 方法返回 true，并为 apply 方法返回一个值。

3. 使用 Spring Boot 进行集成测试

Spring Boot 提供了几个注解来帮助测试。第一个是@SpringBootTest，该注解将使测试成为 Spring Boot 驱动的一个测试。这意味着测试上下文框架将搜索用@SpringBootApplication 注解的类(如果没有传递特定的配置)，并将实际上使用该类启动应用程序。

```
@RunWith(SpringRunner.class)
@SpringBootTest(classes = CalculatorApplication.class)
public class CalculatorApplicationTests {

  @Autowired
  private Calculator calculator;

  @Test(expected = IllegalArgumentException.class)
  public void doingDivisionShouldFail() {
     calculator.calculate(12,13, '/');
  }
}
```

上述测试将启动 CalculatorApplication 应用程序，并注入完全配置好的 Calculator 实例。接着编写一个测试，测试的内容是计算器无法处理的情况，并在代码中指定测试期望引发的异常。

在进行计算时，输出将打印到控制台。使用 JUnit 规则中的 OutputCapture 规则，我们还可以编写一个执行成功的测试用例，并测试打印的输出。

```
@RunWith(SpringRunner.class)
@SpringBootTest(classes = CalculatorApplication.class)
public class CalculatorApplicationTests {

  @Rule
  public OutputCapture capture = new OutputCapture();

  @Autowired
  private Calculator calculator;

  @Test
  public void doingMultiplicationShouldSucceed() {
    calculator.calculate(12,13, '*');
    capture.expect(Matchers.containsString("12 * 13 = 156"));
  }

  @Test(expected = IllegalArgumentException.class)
  public void doingDivisionShouldFail() {
      calculator.calculate(12,13, '/');
  }
}
```

@Rule 注解将配置指定的 JUnit 规则，在本例中是 OutputCapture 规则，它已集成在 Spring Boot 中。该规则截取 System.out 和 System.err 输出流，以便在这两个流上生成的输出中写入断言。在这里，我们做一个乘法运算，输出中应该打印相关的断言。

4. 使用 Spring Boot 和模拟进行集成测试

Spring Boot 使得在应用程序上下文中用模拟来替换 bean 变得很容易。为此，Spring Boot 引入了 @MockBean 注解。

```
@MockBean
private Calculator calculator
```

上述代码将用一个模拟实例替换整个计算器，为了能够使用这个模拟实例，需要使用常规的 Mockito 方式定义其行为。当有多个特定类型的 bean 时，需要指定要替换的 bean 的名称。

```
@MockBean(name ="addition")
private Operation mockOperation;
```

上述代码将用一个模拟实例替换普通的 Addition bean，接下来，可以在该实例上注册模拟行为。当找不到具有该名称的 bean 时，模拟 bean 将注册为该 bean 的新实例。

```
@MockBean(name ="division")
private Operation mockOperation;
```

上述代码不是替换现有的 bean，而是将一个新的 bean 添加到应用程序上下文中，结果是将一个额外的操作添加到 Calculator 可以处理的操作集合中。可以使用 Spring Test 框架中的 ReflectionTestUtils 帮助器类来测试这一点。

```
@Test
public void calculatorShouldHave3Operations() {
  Object operations =
      ReflectionTestUtils.getField(calculator, "operations");

  assertThat((Collection) operations).hasSize(3);
}
```

上述代码将通过反射获取 operations 字段，并断言集合大小为 3。删除@MockBean 注解(或将 name 属性的值修改为 addition)时，此测试将失败，因为现在只注册了两个操作。

使用模拟实例的方式是很简单的：

```
@Test
public void mockDivision() {
  when(mockOperation.handles('/')).thenReturn(true);
  when(mockOperation.apply(14, 7)).thenReturn(2);

  calculator.calculate(14,7, '/');
  capture.expect(Matchers.containsString("14 / 7 = 2"));
}
```

在上述代码中，我们指示 Mockito 在测试除法(/)时返回 true，在调用 apply 方法时返回除法计算的值。接下来调用 calculate 方法，并断言计算的结果与测试的期望值一致。

2.4 配置日志

2.4.1 问题

你希望为某些日志记录器配置日志级别。

2.4.2 解决方案

通过 Spring Boot，你可以对日志框架及其配置信息进行配置。

2.4.3 工作原理

Spring Boot 为支持的日志工具(Logback[5]、Log4j 2[6] 和 Java Util Logging)提供默认配置。除了默认配置之外，它还支持通过常规的 application.properties 文件配置日志级别，以及支持指定日志模式和指定写入日志文件的目录。

Spring Boot 使用 SLF4J 作为日志 API。在编写组件时，应该使用这些接口来编写日志记录。这种方式允许你选择使用哪个日志框架。

1. 配置日志记录

使用日志框架的一个一般事项是对框架的某些部分启用或禁用日志记录。使用 Spring Boot，你可以通过在 application.properties 文件中添加一些属性配置来实现这一点。这些属性配置需要以 logging.level 作为前缀，后跟日志记录器的名称，最后是你想要配置的级别。

```
logging.level.org.springframework.web=DEBUG
```

上述属性配置将启用 org.springframework.web 记录器的 DEBUG 日志记录(通常是针对该包和子包中的所有类)。要设置根日志记录器的级别，需要使用 logging.level.root=<level> 进行配置。这将设置日志记录的默认级别。

2. 将日志写入文件

默认情况下，Spring Boot 仅将日志打印到控制台。如果还需要写入文件，则需要指定 logging.file 或 logging.path 属性。第一个属性指定文件名，第二个属性指定路径。日志记录使用的默认文件名是 spring.log，使用的默认目录是 Java 临时目录。

```
logging.file=application.log
logging.path=/var/log
```

使用上述配置，将在/var/log 目录下写入名为 application.log 的日志文件。

---

[5] https://logback.qos.ch.
[6] https://logging.apache.org/log4j/2.x/.

将日志写入文件时，你可能希望防止日志文件溢出系统。可使用 logging.file.max-history 属性指定日志文件的数量(默认值为 0 表示无限制)，使用 logging.file.max-size 属性指定日志文件的大小(默认值为 10MB)。

3. 使用偏好的日志记录工具

默认情况下，Spring Boot 使用 Logback 框架作为日志记录工具。然而，它还支持 Java Util Logging 框架和 Log4j 2 框架。要使用另一个日志框架，必须首先排除默认框架并包含自己的框架。Spring Boot 的 spring-boot-starter-log4j2 依赖项用于包含 Log4j 2 框架的所有必需的依赖项。要排除默认的 Logback 框架，需要向 spring-boot-starter 依赖项添加一个排除规则，排除的内容是引入默认日志记录框架的主要依赖项。

```xml
<dependency>
  <groupId>org.springframework.boot</groupId>
  <artifactId>spring-boot-starter</artifactId>
  <exclusions>
    <exclusion>
      <groupId>org.springframework.boot</groupId>
      <artifactId>spring-boot-starter-logging</artifactId>
    </exclusion>
  </exclusions>
</dependency>
<dependency>
  <groupId>org.springframework.boot</groupId>
  <artifactId>spring-boot-starter-log4j2</artifactId>
</dependency>
```

> **注意：**
> Spring Boot 2 不支持旧的 Log4j 框架，它支持其后续版本 Log4j 2！

2.5 重用现有配置

2.5.1 问题

你有一个已经存在的、未基于 Spring Boot 的应用程序或模块，并且希望 Spring Boot 重用其配置。

2.5.2 解决方案

要导入现有配置，可将@Import 或@ImportResource 注解添加到@Configuration 或@SpringBootApplication 注解的类中以导入配置。

2.5.3 工作原理

在主应用程序类(带有@SpringBootApplication 注解的类)上，添加@Import 或@ImportResource 注解，以便 Spring 加载其他配置文件。

1. 重用已经存在的 XML 配置

找到带有@SpringBootApplication 注解的类，并将@ImportResource 注解添加到该类上。

```
@SpringBootApplication
@ImportResource("classpath:application-context.xml")
public class Application { ... }
```

由于使用了 classpath:前缀，此配置将从类路径指定的位置加载 application-context.xml 文件。如果该文件位于文件系统中的某个位置，则可以使用 file:前缀，例如，file:/var/conf/application-context.xml。

现在，当启动应用程序时，Spring Boot 还将从上述 XML 文件中加载附加配置。

2. 重用已经存在的 Java 配置

找到带有@SpringBootApplication 注解的类，并将@Import 注解添加到该类上。

```
@SpringBootApplication
@Import(ExistingConfiguration.class)
public class Application { ... }
```

@Import 注解负责将上述类添加到配置中。如果希望包括组件扫描中未涉及的包或组件，或者如果已经禁用了@Configuration 注解的类的自动检测，则可能需要这样做。

第 3 章

Spring MVC 基础特性

当 Spring Boot 在类路径上找到 Spring MVC 类时，将自动配置 Web 应用。它还将启动一个内嵌的服务器(默认情况下，将启动内嵌的 Tomcat[1] 服务器)。

3.1 开始使用 Spring MVC

3.1.1 问题

你希望使用 Spring Boot 开发 Spring MVC 应用程序。

3.1.2 解决方案

Spring Boot 将自动配置 Spring MVC 所需的所有组件。为了实现这一点，Spring Boot 需要能够在其类路径上检测到 Spring MVC 类。为此，需要添加 spring-boot-starter-web 作为依赖项。

3.1.3 工作原理

在项目中添加 spring-boot-starter-web 依赖项。

```xml
<dependency>
  <groupId>org.springframework.boot</groupId>
  <artifactId>spring-boot-starter-web</artifactId>
</dependency>
```

上述配置将为 Spring MVC 添加所需的依赖项。由于 Spring Boot 可以检测到 Spring MVC 类，因此它将配置额外的信息来设置 DispatcherServlet。它还将添加启动内嵌的 Tomcat 服务器所需的所有 JAR 文件。

---

[1] https://tomcat.apache.org.

```
package com.apress.springbootrecipes.hello;

import org.springframework.boot.SpringApplication;
import org.springframework.boot.autoconfigure.SpringBootApplication;

@SpringBootApplication
public class HelloWorldApplication {

  public static void main(String[] args) {
    SpringApplication.run(HelloWorldApplication.class, args);
  }
}
```

这几行代码足以启动内嵌的 Tomcat 服务器并准备好预配置的 Spring MVC 设置。当启动应用程序时，你将看到类似于图 3-1 中的输出。

图 3-1　启动应用程序输出的日志记录

在启动 HelloWorldApplication 时发生了如下事情：
(1) 在端口 8080(默认端口)上启动内嵌的 Tomcat 服务器。
(2) 注册和启用几个默认的 Servlet Filter(如表 3-1 所示)。
(3) 为.css、.js 和 favicon.ico 等文件设置静态资源处理方法。
(4) 与 WebJars[2] 进行集成。
(5) 设置基本的错误处理功能。
(6) 使用所需的组件(例如，ViewResolvers、I18N 等)对 DispatcherServlet 进行预配置。

---

2 https://www.webjars.org.

表 3-1　自动注册的 Servlet Filter 清单

过滤器	说明
CharacterEncodingFilter	默认情况下将强制使用 UTF-8 编码,可以通过设置 spring.http.encoding.charset 属性进行配置。可以通过将 spring.http.encoding.enabled 设置为 false 禁用该属性
HiddenHttpMethodFilter	允许使用名为_method 的隐藏表单字段指定实际的 HTTP 方法。可以通过将 spring.mvc.hiddenmethod.filter.enabled 设置为 false 禁用该属性
FormContentFilter	封装 Put、Patch 和 Delete 请求,以便对这些消息进行绑定处理。可以通过将 spring.mvc.formcontent.filter.enabled 设置为 false 禁用该属性
RequestContextFilter	将当前请求公开到当前线程,以便即使在非 Spring MVC 应用程序(如 Jersey)中也可以使用 RequestContextHolder 和 LocaleContextHolder 进行处理

在当前状态下,HelloWorldApplication 除了启动服务器之外什么都不做。让我们添加一个控制器程序来返回一些信息。

```
package com.apress.springbootrecipes.hello;

import org.springframework.web.bind.annotation.GetMapping;
import org.springframework.web.bind.annotation.RestController;

@RestController
public class HelloWorldController {

  @GetMapping("/")
  public String hello() {
    return "Hello World, from Spring Boot 2!";
  }
}
```

HelloWorldController 类将注册在根目录/下,调用该类的方法时将返回短语"Hello World, from Spring Boot 2!"。@RestController 注解指示这是一个@Controller 类,因此将被 Spring Boot 检测到。此外,它还将@ResponseBody 注解添加到所有处理请求的方法上,指示它应该将结果发送到客户端。@GetMapping 注解将 hello 方法映射到根目录,以处理每一个到达根目录的 GET 请求。我们也可以使用@RequestMapping(value="/",method=RequestMethod.GET)注解实现相同的功能。

重新启动 HelloWorldApplication 时,Spring Boot 将检测并处理 HelloWorldController 类。现在,当使用 curl 或 http 访问 http://localhost:8080/时,结果应该如图 3-2 所示。

```
[→ ~ http http://localhost:8080/
HTTP/1.1 200
Content-Length: 32
Content-Type: text/plain;charset=UTF-8
Date: Wed, 28 Mar 2018 09:49:28 GMT

Hello World, from Spring Boot 2!

→ ~
```

图 3-2　控制器程序的输出

测试

现在应用程序正在运行并返回结果，是时候对控制器程序进行测试了(理想情况下，你应该先编写测试程序！)。Spring 已经有了一些令人印象深刻的测试特性，而 Spring Boot 增加了更多的测试特性。使用 Spring Boot 测试控制器程序相当容易。

```
package com.apress.springbootrecipes.hello;

import org.junit.Test;
import org.junit.runner.RunWith;
import org.springframework.beans.factory.annotation.Autowired;
import org.springframework.boot.test.autoconfigure.web.servlet.WebMvcTest;
import org.springframework.http.MediaType;
import org.springframework.test.context.junit4.SpringRunner;
import org.springframework.test.web.servlet.MockMvc;
import org.springframework.test.web.servlet.request.MockMvcRequestBuilders;

import static org.springframework.test.web.servlet.result.MockMvcResultMatchers.content;
import static org.springframework.test.web.servlet.result.MockMvcResultMatchers.status;

@RunWith(SpringRunner.class)
@WebMvcTest(HelloWorldController.class)
public class HelloWorldControllerTest {
```

```
    @Autowired
    private MockMvc mockMvc;

    @Test
    public void testHelloWorldController() throws Exception {
      mockMvc.perform(MockMvcRequestBuilders.get("/"))
        .andExpect(status().isOk())
        .andExpect(content().string("Hello World, from Spring Boot 2!"))
        .andExpect(content().contentTypeCompatibleWith(MediaType.TEXT_PLAIN));
    }
}
```

需要使用@RunWith(SpringRunner.class)注解来指示JUnit运行这个特定的程序。这个特殊程序将 Spring 测试框架与 JUnit 集成在一起。@WebMvcTest 注解指示 Spring Test 框架设置用于测试该特定控制器程序的应用程序上下文。它将只使用与 Web 相关的 bean(如使用@Controller、@ControllerAdvice 注解的方法等)启动一个最小化的 Spring Boot 应用程序。此外，它还将预配置 Spring Test Mock MVC 支持，然后就可以自动选择相关的类或方法了。

Spring Test Mock MVC 可以用来模拟向控制器程序发送 HTTP 请求，并对结果使用断言进行验证。上述代码中，我们在根目录/下调用 GET，并期望响应的状态码是 HTTP 200(表示正确的响应)，响应的内容是 Hello World, from Spring Boot 2!纯文本消息。

3.2 使用 Spring MVC 公开 REST 资源

3.2.1 问题

你希望使用 Spring MVC 来公开基于 REST 的资源。

3.2.2 解决方案

你需要 JSON 库来执行 JSON 转换(也可以使用 XML 和其他格式，因为内容协商[3]是 REST 的一部分)。在本书中，我们将使用 Jackson[4]库来处理 JSON 转换。

---

3 https://www.ics.uci.edu/~fielding/pubs/dissertation/evaluation.htm#sec_6_3_2_7.
4 https://github.com/FasterXML/jackson.

3.2.3 工作原理

假设你正在为一个图书馆开发网站，需要开发一个 REST API，以便能够查看图书列表和搜索图书。

spring-boot-starter-web 依赖项(参见 3.1 节)已经默认包含所需的 Jackson 库。

```xml
<dependency>
  <groupId>org.springframework.boot</groupId>
  <artifactId>spring-boot-starter-web</artifactId>
</dependency>
```

> ■ 注意：
> 也可以使用 Google GSON 库，只需要使用适当的 GSON 依赖项即可。

由于你正在为图书馆创建应用程序，它很可能需要对图书进行管理，因此让我们创建一个 Book 类。

```java
package com.apress.springbootrecipes.library;

import java.util.*;

public class Book {

  private String isbn;
  private String title;
  private List<String> authors = new ArrayList<>();

  public Book() {}

  public Book(String isbn, String title, String... authors) {
    this.isbn = isbn;
    this.title = title;
    this.authors.addAll(Arrays.asList(authors));
  }

  public String getIsbn() {
      return isbn;
  }
```

```java
    public void setIsbn(String isbn) {
        this.isbn = isbn;
    }

    public String getTitle() {
        return title;
    }

    public void setTitle(String title) {
        this.title = title;
    }

    public void setAuthors(List<String> authors) {
        this.authors = authors;
    }

public List<String> getAuthors() {
    return Collections.unmodifiableList(authors);
}
@Override
public boolean equals(Object o) {
 if (this == o) return true;
 if (o == null || getClass() != o.getClass()) return false;
 Book book = (Book) o;
 return Objects.equals(isbn, book.isbn);
}

@Override
public int hashCode() {
    return Objects.hash(isbn);
}

@Override
public String toString() {
 return String.format("Book [isbn=%s, title=%s, authors=%s]",
                    this.isbn, this.title, this.authors);
 }
}
```

一本书是由它的 ISBN 定义的，它有一个书名和一个或多个作者。

你还需要一个服务来处理图书馆中的图书。为此，先定义 BookService 接口，并实现该接口。

```java
package com.apress.springbootrecipes.library;

import java.util.Optional;

public interface BookService {

  Iterable<Book> findAll();
  Book create(Book book);
  Optional<Book> find(String isbn);
}
```

为简化起见，我们将在内存中管理图书馆的图书，不涉及数据库相关的技术。

```java
package com.apress.springbootrecipes.library;

import org.springframework.stereotype.Service;
import java.util.Map;
import java.util.Optional;
import java.util.concurrent.ConcurrentHashMap;

@Service
class InMemoryBookService implements BookService {

  private final Map<String, Book> books = new ConcurrentHashMap<>();

  @Override
  public Iterable<Book> findAll() {
     return books.values();
  }

  @Override
  public Book create(Book book) {
    books.put(book.getIsbn(), book);
    return book;
  }
```

```
    @Override
    public Optional<Book> find(String isbn) {
        return Optional.ofNullable(books.get(isbn));
    }
}
```

该服务已经用@service 进行了注解，因此 Spring Boot 可以检测到它并创建它的实例。

```
package com.apress.springbootrecipes.library;

import org.springframework.boot.ApplicationRunner;
import org.springframework.boot.SpringApplication;
import org.springframework.boot.autoconfigure.SpringBootApplication;
import org.springframework.context.annotation.Bean;

@SpringBootApplication
public class LibraryApplication {

  public static void main(String[] args) {
      SpringApplication.run(LibraryApplication.class, args);
  }

  @Bean
  public ApplicationRunner booksInitializer(BookService bookService) {
    return args -> {
      bookService.create(
        new Book("9780061120084", "To Kill a Mockingbird", "Harper
                Lee"));
      bookService.create(
        new Book("9780451524935", "1984", "George Orwell"));
      bookService.create(
        new Book("9780618260300", "The Hobbit", "J.R.R. Tolkien"));
    };
  }
}
```

LibraryApplication 类将检测所有的类并启动服务器。一开始，它将预先登记三本书，以便在后续的示例中对图书馆的图书进行操作演示。

要将 Book 类作为 REST 资源公开，可以创建一个 BookController 类，并使用 @RestController 对其进行注解。Spring Boot 将检测这个类并创建它的一个实例。使用@RequestMapping(或者@GetMapping、@PostMapping)来注解为处理传入的请求而编写的方法。

■ **注意:**
除了使用@RestController注解之外，还可以使用@Controller注解并在每个处理请求的方法上添加@ResponseBody注解。使用@RestController注解将隐式地把@ResponseBody注解添加到处理请求的方法上。

```java
package com.apress.springbootrecipes.library.rest;

import com.apress.springbootrecipes.library.Book;
import com.apress.springbootrecipes.library.BookService;
import org.springframework.http.ResponseEntity;
import org.springframework.web.bind.annotation.*;
import org.springframework.web.util.UriComponentsBuilder;

import java.net.URI;

@RestController
@RequestMapping("/books")
public class BookController {

    private final BookService bookService;

    public BookController(BookService bookService) {
        this.bookService = bookService;
    }

    @GetMapping
    public Iterable<Book> list() {
        return bookService.findAll();
    }

    @GetMapping("/{isbn}"
    public ResponseEntity<Book> get(@PathVariable("isbn") String isbn) {
     return bookService.find(isbn)
        .map(ResponseEntity::ok)
        .orElse(ResponseEntity.notFound().build());
    }
```

```
@PostMapping
public Book create(@RequestBody Book book,
                   UriComponentsBuilder uriBuilder) {
  Book created = bookService.create(book);
  URI newBookUri = uriBuilder.path("/books/{isbn}").build
                   (created.getIsbn());
  return ResponseEntity.created(newBookUri).body(created);
  }
}
```

由于使用@RequestMapping("/books")对类进行注解，控制器程序将被映射到/books 路径。在/books 路径上调用 GET 请求时将触发应用程序的 list 方法。当使用/books/<isbn>触发 GET 请求时，应用程序将调用上述代码中的 get 方法并返回单本书的信息，或者在无法找到对应的书时返回 404 响应状态码。最后，可以在/books路径上调用 POST 请求将图书添加到图书馆。接收到 POST 请求后，应用程序将调用 create 方法并将传入请求的主体内容转换为图书的记录信息。

应用程序启动后，可以使用 HTTPie[5] 或 curl[6] 来检索图书。当使用 HTTPie 访问 http://localhost:8080/books 时，你应该会看到类似于图 3-3 的输出。

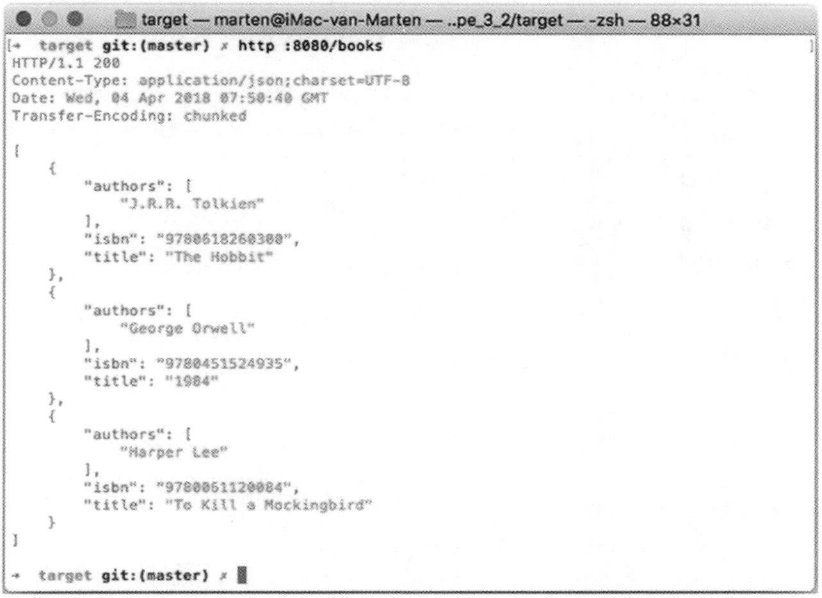

图 3-3　以 JSON 格式输出的图书列表

---

5 https://httpie.org.

6 https://curl.haxx.se.

对 http://localhost:8080/books/9780451524935 触发 GET 请求将返回一本书的信息，在上述示例中是 George Orwell 撰写的《1984》。使用一个未知的 ISBN 发起请求将导致返回的状态码为 404。

当发出一个 POST 请求时，可以在图书列表中添加一本新书。

```
http POST :8080/books \
  title="The Lord of the Rings" \
  isbn="9780618640157" \
  authors:='["J.R.R. Tolkien"]'
```

上述 POST 请求的结果是，如果应用程序处理正确，为图书馆添加一本新的图书，并生成获取该图书信息的路径。现在，当你获取图书列表时，它应该包含四本书而不是一开始添加的三本书。

在处理上述 POST 请求时，HTTPie 将参数转换为一个 JSON 请求主体，然后由 Jackson 库读取，并将其转换为一个 Book 实例，如下所示。

```
{
  "title": "The Lord of the Rings",
  "isbn": "9780618640157",
  "authors": ["J.R.R. Tolkien"]
}
```

默认情况下，Jackson 将使用 getter 和 setter 方法将 JSON 映射到一个对象。具体情况是，应用程序使用默认的无参构造函数创建一个新的 Book 实例，并通过不同的 setter 方法设置所有属性。例如，对于 title 属性，使用 setTitle 方法设置等。

测试@RestController 注解的类

由于希望确保控制器程序执行了它应该做的事情，因此编写一个测试来验证控制器程序的正确行为。

```
Package com.apress.springbootrecipes.library.rest;

import com.apress.springbootrecipes.library.Book;
import com.apress.springbootrecipes.library.BookService;
import org.junit.Test;
import org.junit.runner.RunWith;
import org.springframework.beans.factory.annotation.Autowired;
import org.springframework.boot.test.autoconfigure.web.servlet
.WebMvcTest;
import org.springframework.boot.test.mock.mockito.MockBean;
import org.springframework.test.context.junit4.SpringRunner;
```

```java
import org.springframework.test.web.servlet.MockMvc;
import org.springframework.test.web.servlet.result
.MockMvcResultMatchers;

import java.util.Arrays;
import java.util.Optional;

import static org.hamcrest.Matchers.*;
import static org.mockito.ArgumentMatchers.anyString;
import static org.mockito.Mockito.when;
import static org.springframework.test.web.servlet.request.
MockMvcRequestBuilders.get;
import static org.springframework.test.web.servlet.result.
MockMvcResultMatchers.status;

@RunWith(SpringRunner.class)
@WebMvcTest(BookController.class)
public class BookControllerTest {

    @Autowired
    private MockMvc mockMvc;

    @MockBean
    private BookService bookService;

    @Test
    public void shouldReturnListOfBooks() throws Exception {

        when(bookService.findAll()).thenReturn(Arrays.asList(
            new Book("123","Spring 5 Recipes","Marten Deinum","Josh Long"),
            new Book("321","Pro Spring MVC","Marten Deinum","Colin Yates")));

        mockMvc.perform(get("/books"))
            .andExpect(status().isOk())
            .andExpect(jsonPath("$", hasSize(2)))
            .andExpect(jsonPath("$[*].isbn", containsInAnyOrder("123",
                    "321")))
            .andExpect(jsonPath("$[*].title",
```

```
            containsInAnyOrder("Spring 5 Recipes", "Pro Spring MVC")));
    }

    @Test
    public void shouldReturn404WhenBookNotFound() throws Exception {

      when(bookService.find(anyString())).thenReturn(Optional.empty());

      mockMvc.perform(get("/books/123")).andExpect(status()
                        .isNotFound());
    }

    @Test
    public void shouldReturnBookWhenFound() throws Exception {

      when(bookService.find(anyString())).thenReturn(
        Optional.of(
          new Book("123","Spring 5 Recipes","Marten Deinum","Josh Long")));

      mockMvc.perform(get("/books/123"))
        .andExpect(status().isOk())
        .andExpect(jsonPath("$.isbn", equalTo("123")))
        .andExpect(jsonPath("$.title", equalTo("Spring 5 Recipes")));
    }
}
```

上述测试使用@WebMvcTest 注解创建基于 MockMvc 的测试,并将创建一个最小化的 Spring Boot 应用程序,以便能够运行控制器程序。控制器程序需要一个 BookService 类的实例,所以我们让框架通过@MockBean 注解为它创建一个模拟实例。在不同的测试方法中,我们模拟不同的预期行为,例如返回书的列表、返回空的 Optional 实例等。

■ 注意:
Spring Boot使用Mockito[7]的@MockBean注解来创建模拟实例。

而且,上述测试使用了 JsonPath[8] 库,这样就可以使用表达式来测试 JSON 结果。

---

7 https://site.mockito.org.
8 https://github.com/json-path/JsonPath.

JsonPath 在 JSON 中的作用相当于 XPath 在 XML 中的作用。

3.3 在 Spring Boot 中使用 Thymeleaf 模板

3.3.1 问题

你希望使用 Thymeleaf 模板来呈现应用程序的页面。

3.3.2 解决方案

添加 Thymeleaf 的依赖项，并创建一个常规的@Controller 注解类来确定视图并填充模型。

3.3.3 工作原理

作为开始，首先需要在项目中添加 spring-boot-starter-thymeleaf 依赖项，以获取使用 Thymeleaf 模板[9]所需的依赖项。

```
<dependency>
  <groupId>org.springframework.boot</groupId>
  <artifactId>spring-boot-starter-thymeleaf</artifactId>
</dependency>
```

添加这个依赖项之后，Spring Boot 将加载 Thymeleaf 库和 Thymeleaf Spring Dialect 库，以便这两个库能充分集成。由于这两个库的存在，Spring Boot 将自动配置 ThymeleafViewResolver。

ThymeleafViewResolver 需要 ThymeleafItemplateEngine 来解析和呈现视图。Spring Boot 将使用 SpringDialect 来配置一个特殊的 SpringTemplateEngine，这样就可以在 Thymeleaf 页面中使用 SpEL 表达式了。

为配置 Thymeleaf，Spring Boot 在 spring.thymeleaf 名称空间中公开了表 3-2 中的几个属性。

表 3-2 配置 Thymeleaf 的属性

属性	说明
spring.thymeleaf.prefix	ViewResolver 使用的前缀，默认值为 classpath:/templates/
spring.thymeleaf.suffix	ViewResolver 使用的后缀，默认值为.html
spring.thymeleaf.encoding	模板的编码，默认值为 UTF-8
spring.thymeleaf.check-template	在呈现之前检查模板是否存在，默认值为 true

---

9 https://www.thymeleaf.org。

(续表)

属性	说明
spring.thymeleaf.check-template-location	检查模板的目录是否存在，默认值为 true
spring.thymeleaf.mode	要使用的 Thymeleaf TemplateMode，默认为 HTML
spring.thymeleaf.cache	是否缓存解析后的模板，默认值为 true
spring.thymeleaf.template-resolver-order	ViewResolver 的呈现优先级，默认为 1
spring.thymeleaf.view-names	可以用此 ViewResolver 解析的视图名称，多个名称之间用逗号分隔
spring.thymeleaf.excluded-view-names	被排除在解析之外的视图名称，多个名称之间用逗号分隔
spring.thymeleaf.enabled	是否启用 Thymeleaf 模板，默认值为 true
spring.thymeleaf.enable-spring-el-compiler	是否启用 SpEL 表达式的编译，默认值为 false
spring.thymeleaf.servlet.content-type	编写 HTTP 响应所使用的 Content-Type 值，默认为 text/html

1. 添加索引页

首先向应用程序添加一个索引页面。在 src/main/resources/templates 目录(默认位置)中创建 index.html 文件。

```html
<!DOCTYPE html>
<html xmlns:th="http://www.thymeleaf.org">
<head>
  <meta charset="UTF-8">
  <title>Spring Boot Recipes - Library</title>
</head>
<body>

<h1>Library</h1>

<a th:href="@{/books.html}" href="#">List of books</a>

</body>
</html>
```

这只是一个基本的 HTML5 的网页，其中增加了一点 Thymeleaf 模板的内容。首先增加的是 xmlns:th="http://www.thymeleaf.org"，用于启用 Thymeleaf 的名称空间。

接下来通过 th:href 属性在链接中使用名称空间。@{/books.html}将由 Thymeleaf 扩展为一个适当的链接，并具体放置在链接的 href 属性中。

现在，当运行应用程序并访问主页(http://localhost:8080/)时，你应该会看到一个页面，该页面链接到图书概览，如图 3-4 所示。

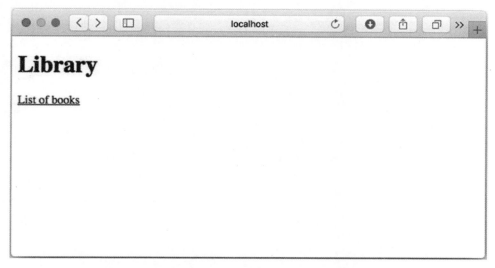

图 3-4　图书馆的索引页

2. 添加控制器和视图

单击索引页中提供的链接时，我们希望显示一个页面，其中显示了图书馆中可用图书的列表，如图 3-5 所示。为此，需要添加一个类和一个视图：首先是可以处理请求并准备模型的控制器类，其次是呈现图书列表的视图。

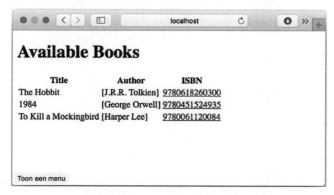

图 3-5　图书列表页面

让我们添加一个控制器，它将用图书的列表填充模型，并选择要呈现的视图的名称。控制器是一个带有 @Controller 注解的类，它包含请求处理方法(用

@RequestMapping 注解的方法，或是本书中使用@GetMapping 注解的方法，它是一个专用的@RequestMapping 注解）。

```
package com.apress.springbootrecipes.library.web;

@Controller
public class BookController {

  private final BookService bookService;

  public BookController(BookService bookService) {
    this.bookService = bookService;
  }

  @GetMapping("/books.html")
  public String all(Model model) {
     model.addAttribute("books", bookService.findAll());
     return "books/list";
  }
}
```

BookController 类需要从 BookService 类获取要显示的图书列表。上述代码中的 all 方法使用 org.springframework.ui.Model 作为参数，这样就可以将图书列表放置到模型中。请求处理方法可以有不同的参数 [10]，其中之一是 Model 类。在 all 方法中，我们使用 BookService 从数据存储中检索所有图书，并使用 model.addAttribute 将其添加到模型中。现在在模板中可以通过键值 books 获取图书列表，接下来的代码将进行演示。

最后，我们返回用于呈现 books/list 的视图的名称。这个名字被传递到 ThymeleafViewResolver，并组成一个路径，即 classpath:/templates/books/list.html。

在添加了控制器和请求处理方法之后，需要创建视图。在 src/main/templates/books 目录中创建 list.html 文件。

```
<!DOCTYPE html>
<html xmlns:th="http://www.thymeleaf.org">
<head lang="en">
  <meta charset="UTF-8">
  <title>Library - Available Books</title>
  <meta http-equiv="Content-Type" content="text/html;
```

---

10 https://docs.spring.io/spring/docs/current/spring-framework-reference/web.html#mvc-ann-arguments.

```html
                         charset=UTF-8"/>
    </head>
    <body>
      <h1>Available Books</h1>
      <table>
        <thead>
          <tr>
              <th>Title</th>
          <th>Author</th>
          <th>ISBN</th>
          </tr>
        </thead>
      <tbody>
        <tr th:each="book : ${books}">
          <td th:text="${book.title}">Title</td>
          <td th:text="${book.authors}">Authors</td>
          <td>
            <a th:href="@{/books.html(isbn=${book.isbn})}" href="#"
               th:text="${book.isbn}">1234567890123</a>
          </td>
        </tr>
      </tbody>
      </table>
    </body>
    </html>
```

这同样是一个使用了 Thymeleaf 语法的 HTML5 页面。该页面使用 th:each 表达式呈现图书列表。它将通过模型中的 books 关键字获取所有图书，并在表格中为每本书创建一行。行中的每一列都包含使用 th:text 表达式的相关代码，它将打印书的书名、作者和 ISBN，表中的最后一列包含一个指向图书详细信息的链接。代码中使用 th:href 表达式构造一个 URL。注意，我们通过使用 th:href="@{/books.html(isbn=${book.isbn})}"表达式将 isbn 作为请求参数添加到链接中。

当启动应用程序并点击索引页上的链接时，你应该会看到一个显示图书馆已有图书的页面，如图 3-5 所示。

3. 添加详细信息页面

最后，当单击表格中的 ISBN 时，你希望显示一个包含图书详细信息的页面。该链接包含一个名为 isbn 的请求参数，我们可以在控制器中使用该参数进行检索以找到图书。在方法的参数上添加@RequestParam 注解后，调用该方法时，Spring Boot

会为应用程序检索该参数,下面的代码将进行演示。

下面的方法将处理 GET 请求,将请求参数映射到方法参数,并包含该模型,这样我们就可以在模型中添加图书的信息。

```java
@GetMapping(value = "/books.html", params = "isbn")
public String get(@RequestParam("isbn") String isbn, Model model) {

  bookService.find(isbn)
        .ifPresent(book -> model.addAttribute("book", book));

  return "books/details";
}
```

控制器将呈现 books/details 页面。将 details.html 添加到 src/main/resources/templates/books 目录中。

```html
<!DOCTYPE html>
<html xmlns:th="http://www.thymeleaf.org">
<head lang="en">
  <meta charset="UTF-8">
  <title>Library - Available Books</title>
  <meta http-equiv="Content-Type" content="text/html;
        charset=UTF-8"/>
</head>
<body>
  <div th:if="${book != null}">
    <div>
        <div th:text="${book.title}">Title</div>
        <div th:text="${book.authors}">authors</div>
        <div th:text="${book.isbn}">ISBN</div>
    </div>
  </div>

  <div th:if="${book} == null">
    <h1 th:text="'No book found with ISBN: ' + ${param.isbn}">Not Found</h1>
  </div>
</body>
</html>
```

上述 HTML5 页面中的 Thymeleaf 模板包括两个块，页面将根据查询结果显示其中的一个。如果已经找到了图书，它将显示图书的详细信息；否则，它将显示一条消息，说明未找到相关的图书。这个逻辑可以通过使用 th:if 表达式来实现。未找到相关图书时，展现的信息中使用的 isbn 是通过使用 param 前缀从请求参数中获取的。即，上述代码中的${param.isbn}将获取请求消息中的 isbn 参数。

3.4 处理异常

3.4.1 问题

你希望自定义默认的白标签错误页面，该页面由 Spring Boot 显示。

3.4.2 解决方案

添加一个额外的 error.html 作为自定义错误页面，或者为特定的 HTTP 错误代码添加自定义的错误页面。例如，可以为返回状态码 404 添加 404.html，用于显示返回状态码为 404 时的错误信息。

3.4.3 工作原理

默认情况下，Spring Boot 启用了错误处理功能，并会显示一个默认的错误页面。可以通过将 server.error.whitelabel.enabled 属性设置为 false 来完全禁用默认的错误页面。当禁用时，异常将由 Servlet 容器来处理，而不是由 Spring 和 Spring Boot 提供的通用异常处理机制进行处理。

还有其他一些属性可以用来配置白标签错误页面，这些属性主要用于配置在模型中包含哪些内容，以便将这些内容呈现给用户。表 3-3 列出了这些属性。

表 3-3 错误处理属性

属性	说明
server.error.whitelabel.enabled	是否启用白标签错误页，默认为 true
server.error.path	错误页面的路径，默认路径为/error
server.error.include-exception	模型中是否包含异常的名称，默认为 false
server.error.include-stacktrace	模型中是否包含堆栈跟踪信息，默认为 never

首先，在 BookController 类中添加一个方法，该方法强制抛出一个异常。

```
@GetMapping("/books/500")
public void error() {
    throw new NullPointerException("Dummy NullPointerException.");
}
```

上述方法将抛出一个异常，因此，在访问 http://localhost:8080/books/500 时，浏览器将显示白标签错误页面，如图 3-6 所示。

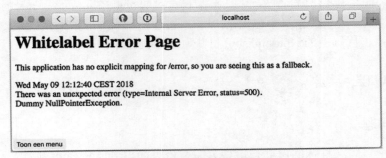

图 3-6 默认的错误页面

如果找不到其他错误页面，则显示此页面。要覆盖该页面，可以在 src/main/resources/templates 目录中添加 error.html。

```
<!DOCTYPE html>
<html xmlns:th="http://www.thymeleaf.org">
<head>
  <meta charset="UTF-8">
  <title>Spring Boot Recipes - Library</title>
</head>
<body>
<h1>Oops something went wrong, we don't know what but we are going to work on it!</h1>

<div>
  <div>
      <span><strong>Status</strong></span>
      <span th:text="${status}"></span>
  </div>
  <div>
      <span><strong>Error</strong></span>
      <span th:text="${error}"></span>
  </div>
  <div>
      <span><strong>Message</strong></span>
      <span th:text="${message}"></span>
  </div>
  <div th:if="${exception != null}">
```

```html
        <span><strong>Exception</strong></span>
        <span th:text="${exception}"></span>
    </div>
    <div th:if="${trace != null}">
        <h3>Stacktrace</h3>
        <span th:text="${trace}"></span>
    </div>
  </div>
</body>
</html>
```

现在，当应用程序启动并发生异常时，这个自定义错误页面就会显示出来，如图 3-7 所示。该页面将通过你选择的视图模板呈现(本例中使用的模板是 Thymeleaf)。

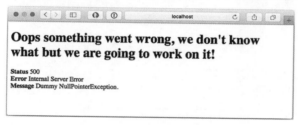

图 3-7　自定义的错误页面

如果你现在将 server.error.include-exception 属性设置为 true，并将 server.error.include-stacktrace 属性设置为 always，那么自定义的错误页面还将包含异常的类名和堆栈跟踪信息，如图 3-8 所示。

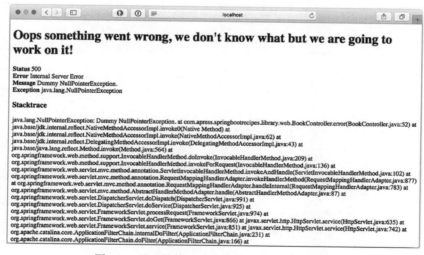

图 3-8　显示堆栈跟踪信息的自定义错误页面

除了提供自定义的通用错误页面之外，你还可以为特定的 HTTP 状态码添加错误页面。这可以通过在 src/main/resources/templates/error 目录中添加 <http-status>.html 来实现，其中 http-status 为相应的 HTTP 状态码。下面将演示为 404 状态码添加 404.html，当通过未知的 URL 访问图书馆时，应用程序将显示该错误页面。

```html
<!DOCTYPE html>
<html xmlns:th="http://www.thymeleaf.org">
<head>
  <meta charset="UTF-8">
  <title>Spring Boot Recipes - Library - Resource Not Found</title>
</head>
<body>
<h1>Oops the page couldn't be located.</h1>
</body>
</html>
```

当导航到应用程序不知道的 URL 时，它会呈现此页面；当触发异常时，它仍然会显示定制的错误页面，如图 3-7 和图 3-8 所示。

■ 提示：
可以为从 400 到 500 范围内的所有HTTP状态码添加自定义的错误页，只需要将状态码作为文件名，后缀为.html即可。

向模型添加属性

默认情况下，Spring Boot 将在模型中包含错误页面的属性，如表 3-4 所示。

表 3-4　模型中默认包含的错误属性

属性	说明
timestamp	错误发生的时间
status	状态代码
error	错误原因
exception	根异常的类名(如果已配置)
message	异常消息
errors	由 BindingResult 产生的任何 ObjectError 对象(仅当使用了消息绑定和/或消息验证时)
trace	异常堆栈跟踪信息(如果已配置)
path	引发异常时的 URL 路径

第 3 章 ■ Spring MVC 基础特性

以上属性及其功能均由 ErrorAttributes 组件提供。Spring Boot 默认使用和配置的组件是 DefaultErrorAttributes。你可以创建自己的 ErrorAttributes 处理程序来创建自定义模型或扩展 DefaultErrorAttributes 以添加其他属性。

```java
package com.apress.springbootrecipes.library;

import org.springframework.boot.web.servlet.error.DefaultErrorAttributes;
import org.springframework.web.context.request.WebRequest;
import java.util.Map;

public class CustomizedErrorAttributes extends DefaultErrorAttributes {

  @Override
  public Map<String, Object> getErrorAttributes(WebRequest webRequest, boolean includeStackTrace) {
    Map<String, Object> errorAttributes =
      super.getErrorAttributes(webRequest, includeStackTrace);
    errorAttributes.put("parameters", webRequest.getParameterMap());
    return errorAttributes;
  }
}
```

CustomizedErrorAttributes 类把原始的请求参数添加到模型的默认属性之后。接下来将这个类配置为 LibraryApplication 类的一个 bean。然后，Spring Boot 将检测并使用这个自定义的异常处理类，而不是配置默认的异常处理类。

```java
@Bean
public CustomizedErrorAttributes errorAttributes() {
    return new CustomizedErrorAttributes();
}
```

最后，你可能需要使用 error.html 中的其他属性。

```html
<div th:if="${parameters != null}">
<h3>Parameters<h3>
<span th:each="parameter :${parameters}">
  <div th:text="${parameter.key} + ' : ' + ${#strings.arrayJoin(parameter.value, ',')}"></div>
```

```
            </span>
        </div>
```

当在 error.html 中包含上述代码时，它将打印模型中可用的参数映射的内容，如图 3-9 所示。

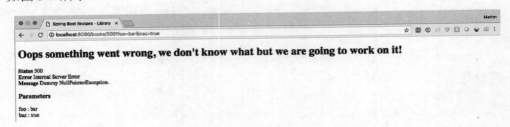

图 3-9　带参数的自定义错误页面

3.5　应用程序国际化

3.5.1　问题

在开发国际化的 Web 应用时，必须根据用户所在的区域显示网页。你不希望为不同区域创建同一页面的不同版本。

3.5.2　解决方案

为了避免为不同的区域创建不同版本的页面，你应该通过外置对区域敏感的文本消息来使你的网页不受用户所在区域的限制。Spring 能够使用消息源解析文本消息，该消息源必须实现 MessageSource 接口。在页面模板中可以使用特殊的标记，或查找消息的内容。

3.5.3　工作原理

当在 src/main/resources 目录(默认位置)中找到 messages.properties 文件时，Spring Boot 会自动配置一个 MessageSource 实例。这个 messages.properties 文件包含了需要在应用程序中使用的默认消息。Spring Boot 将使用来自请求的 Accept-Language 报头来确定当前请求使用哪个区域设置(关于如何更改该参数的内容参见 3.6 小节)。

有一些属性可以改变消息源对丢失的翻译、缓存等作出反应的方式。相关属性的概述如表 3-5 所示。

表 3-5 配置 I18N 的属性

属性	说明
spring.messages.basename	逗号分隔的 basename 列表，默认为 messages
spring.messages.encoding	消息束编码，默认为 UTF-8
spring.messages.always-use-message-format	MessageFormat 是否应用于所有消息，默认为 false
spring.messages.fallback-to-system-locale	如果找不到已检测的区域的资源包，则使用系统区域设置。禁用时，将从默认文件加载默认值，默认为 true
spring.messages.use-code-as-default-message	当找不到任何消息时，可以使用消息代码作为默认消息，而不是抛出 NoSuchMessageException 异常。默认为 false
spring.messages.cache-duration	缓存持续时间，默认为永久保存

■ 提示：

在将应用程序部署到云或其他外部托管方时，将 spring.messages.fallback-to-system-locale 设置为 false 是很有用的。这样，你就可以控制应用程序的默认语言是什么，而不是依赖于你所部署的(不受你控制的)环境。

在 src/main/resources 目录中添加 messages.properties 文件。

```
main.title=Spring Boot Recipes - Library

index.title=Library
index.books.link=List of books

books.list.title=Available Books
books.list.table.title=Title
books.list.table.author=Author
books.list.table.isbn=ISBN
```

现在更改模板以使用这些外置的资源，下面是已修改的 index.html 文件。

```
<!DOCTYPE html>
<html xmlns:th="http://www.thymeleaf.org">
<head>
    <meta charset="UTF-8">
    <title th:text=""#{main.title}">Spring Boot Recipes -
```

```
    Library</title>
    </head>
    <body>

    <h1 th:text="#{index.title}">Library</h1>

    <a th:href="@{/books.html}" href="#"
th:text="#{index.books.link}">List of
    books</a>

    </body>
    </html>
```

对于 Thymeleaf 模板，可以在 th:text 属性中使用 #{..} 表达式。由于 Spring 的自动集成，这将解析来自 MessageSource 的消息。当重新启动应用程序时，它的显示没有任何改变。然而，现在所有的文本都来自 messages.properties 文件。

现在让我们为这个网站的荷兰语翻译添加 messages_nl.properties 文件。

```
main.title=Spring Boot Recipes - Bibliotheek

index.title=Bibliotheek
index.books.link=Lijst van boeken

books.list.title=Beschikbare Boeken
books.list.table.title=Titel
books.list.table.author=Auteur
books.list.table.isbn=ISBN
```

现在，当将 Accept_Language 头的值修改为 Dutch 时，该网站的内容将转换为荷兰语，如图 3-10 所示。

■ 提示：
改变浏览器的语言不是很容易的事情。对于Chrome和Firefox，有一些插件可以让你轻松地切换Accept-Language头的值。

图 3-10　荷兰语主页

3.6　解析用户区域设置

3.6.1　问题

为了使 Web 应用支持国际化，应用程序必须识别每个用户的首选区域设置并根据该区域显示内容。

3.6.2　解决方案

在 Spring MVC 应用程序中，用户的区域设置由区域解析器标识，该解析器必须实现 LocaleResolver 接口。Spring MVC 附带了几个 LocaleResolver 实现，可以根据不同的标准解析区域设置。或者，也可以通过实现此接口来创建自己的自定义区域设置解析器。

Spring Boot 允许设置 spring.mvc.locale-resolver 属性，可以将其设置为 ACCEPT（默认值）或 FIXED。第一种设置将创建一个 AcceptHeaderLocaleResolver 实例，第二种设置将创建一个 FixedLocaleResolver 实例。

还可以通过在 Web 应用上下文中注册一个类型为 LocaleResolver 的 bean 来定义区域设置解析器。必须将区域解析器的 bean 名称设置为 localeResolver，以便 Spring Boot 能够自动检测。

3.6.3　工作原理

Spring MVC 附带了几个默认的 LocaleResolver 接口实现。Spring MVC 还提供

了 HandlerInterceptor 类的子类，这个类允许用户覆盖他们希望使用的语言环境，即 LocaleChangeInterceptor 类。

1. 通过 HTTP 请求的 Accept-Language 头解析区域设置

Spring Boot 注册的默认区域解析器是 AcceptHeaderLocaleResolver。它通过检查 HTTP 请求的 Accept-Language 头来解析区域设置。这个头部字段是由用户的网络浏览器根据底层操作系统的区域设置来设置的。

■ 注意：
AcceptHeaderLocaleResolver 无法更改用户的区域设置，因为它无法修改用户操作系统的区域设置。

```
@Bean
public LocaleResolver localeResolver () {
  return new AcceptHeaderLocaleResolver();
}
```

2. 通过会话属性解析区域设置

解决区域设置的另一个选择是使用 SessionLocaleResolver，它通过检查用户会话中的预定义属性来解析区域设置。如果该会话属性不存在，则此区域设置解析器从 HTTP 请求的 Accept-Language 字段确定默认的区域设置。

```
@Bean
public LocaleResolver localeResolver () {
  SessionLocaleResolver localeResolver = new
    SessionLocaleResolver();
  localeResolver.setDefaultLocale(new Locale("en"));
  return localeResolver;
}
```

在会话属性不存在的情况下，可以为该解析器设置 defaultLocale 属性。请注意，此区域设置解析器可以通过更改存储区域设置的会话属性来更改用户的区域设置。

3. 通过 Cookie 解析区域设置

还可以使用 CookieLocaleResolver 通过检查用户浏览器中的 cookie 来解析区域设置。如果 cookie 不存在，此区域设置解析器将从 HTTP 请求的 Accept-Language 字段确定默认的区域设置。

```
@Bean
public LocaleResolver localeResolver() {
```

```
    return new CookieLocaleResolver();
}
```

可以通过设置 cookieName 和 cookieMaxAge 属性来自定义此区域设置解析器所使用的 cookie。cookieMaxAge 属性指示应保留此 cookie 的秒数。设置该属性的值为-1 表示在浏览器关闭后，该 cookie 将失效。

```
@Bean
public LocaleResolver localeResolver() {
  CookieLocaleResolver cookieLocaleResolver = new
    CookieLocaleResolver();
  cookieLocaleResolver.setCookieName("language");
  cookieLocaleResolver.setCookieMaxAge(3600);
  cookieLocaleResolver.setDefaultLocale(new Locale("en"));
  return cookieLocaleResolver;
}
```

还可以为这个区域设置解析器设置 defaultLocale 属性，以防用户的浏览器中不存在 cookie。此区域设置解析器可以通过更改存储区域设置的 cookie 来更改用户的区域设置。

4. 使用固定的区域设置

FixedLocaleResolver 总是返回相同的固定区域设置。默认情况下，它返回 JVM 默认的区域，但可以通过设置 defaultLocale 属性将返回值配置为其他区域。

```
@Bean
public LocaleResolver localeResolver() {
  FixedLocaleResolver cookieLocaleResolver = new
      FixedLocaleResolver();
  cookieLocaleResolver.setDefaultLocale(new Locale("en"));
  return cookieLocaleResolver;
}
```

■ **注意：**
FixedLocaleResolver 不能更改用户的区域设置，因为顾名思义，它返回的是固定值。

5. 改变用户的区域设置

除了通过显式调用 LocaleResolver.setLocale()来更改用户的区域设置之外，还可以使用 LocaleChangeInterceptor 进行处理器程序映射。该拦截器检测当前 HTTP 请求中是否存在特殊参数。可以使用该拦截器的 paramName 属性定制参数的名称(默认参数名是 locale)。如果当前请求中存在这样的参数，则此拦截器将根据参数值更改用户的区域设置。

为了更改区域设置，必须使用允许更改用户区域设置的 LocaleResolver 对象，例如前面提到的 SessionLocaleResolver 或 CookieLocaleResolver。

```
@Bean
public LocaleResolver localeResolver() {
  return new CookieLocaleResolver();
}
```

为了更改区域设置，请将 LocaleChangeInterceptor 添加为 bean 并将其注册为拦截器。在注册拦截器时，需要使用 WebMvcConfigurer 类的 addInterceptors 方法。

■ 注意：
不一定要将上述拦截器添加到@SpringBootApplication注解的类中，还可以创建一个专用的、用@Configuration注释的类来注册拦截器。注意不要为这个类添加@EnableWebMvc注解，因为这将禁用由Spring Boot进行的自动配置！

```
@SpringBootApplication
public class LibraryApplication implements WebMvcConfigurer {

  @Override
  public void addInterceptors(InterceptorRegistry registry) {
      registry.addInterceptor(localeChangeInterceptor());
  }

  @Bean
  public LocaleChangeInterceptor localeChangeInterceptor() {
    return new LocaleChangeInterceptor();
  }

  @Bean
  public LocaleResolver localeResolver() {
    return new CookieLocaleResolver();
  }
}
```

现在将下面的代码片段添加到 index.html 中：

```html
<h3>Language</h3>
<div>
  <a href="?locale=nl" th:text="#{main.language.nl}">NL</a> |
  <a href="?locale=en" th:text="#{main.language.en}">EN</a>
</div>
```

将如下键值对添加到 messages.properties 文件中。

```
main.language.nl=Dutch
main.language.en=English
```

现在，在选择一种语言时，页面将重新呈现并显示所选择的语言。当继续浏览时，应用程序中的其他页面也将显示所选择的语言。

3.7 选择和配置内嵌的服务器

3.7.1 问题

你希望将 Jetty 用作内嵌的容器，而不是用作默认的 Tomcat 容器。

3.7.2 解决方案

排除 Tomcat 运行时并包含 Jetty 运行时。Spring Boot 将自动检测 Tomcat、Jetty 或 Undertow 是否在类路径上，并根据检测结果相应地配置容器。

3.7.3 工作原理

Spring Boot 支持开箱即用 Tomcat、Jetty 和 Undertow，将它们作为内嵌的 servlet 容器。默认情况下，Spring Boot 使用 Tomcat 作为容器(因为 spring-boot-starter-web 工件的依赖项包含 spring-boot-starter-tomcat)。容器可以使用属性进行配置，其中一些属性适用于所有容器，而其他一些适用于特定容器。全局属性的前缀是 server.或 server.servlet，而容器特定属性的前缀是 server.\<container\>(具体配置属性时，其中的 container 将由 tomcat、jetty 或 undertow 三者中的某一个替代)。

1. 通用配置属性

Spring Boot 提供了几个通用的服务器属性，如表 3-6 所示。

表 3-6 通用服务器属性

属性	说明
server.port	HTTP 服务器端口，默认为 8080
server.address	绑定的 IP 地址，默认为 0.0.0.0(即所有 IP 地址)
server.use-forward-headers	X-Forwarded-*头是否应用于当前请求，默认未设置，使用选定的 servlet 容器中的默认值
server.server-header	服务器响应消息头中发送的值，如果未设置，将使用 servlet 容器的默认值
server.max-http-header-size	HTTP 报头的最大长度，默认值为 0(表示无长度限制)
server.connection-timeout	HTTP 连接器在关闭前等待下一个请求的超时时间。默认值为空，将其留给容器决定；设置为－1 表示连接永不超时
server.http2.enabled	如果当前容器支持，则启用 Http2 支持。默认为 false
server.compression.enabled	是否启用 HTTP 压缩，默认为 false
server.compression.mime-types	需要压缩的 MIME 类型列表，以逗号分隔
server.compression.excluded-user-agents	应禁用压缩的用户代理的列表，以逗号分隔
server.compression.min-response-size	应用压缩的请求的最小长度，默认值为 2048(字节)
server.servlet.context-path	应用程序的主上下文路径，默认情况下作为根应用程序启动
server.servlet.path	主 DispatcherServlet 的路径，默认为根目录
server.servlet.application-display-name	服务器在容器中显示的名称，默认为 application
server.servlet.context-parameters	Servlet 容器的上下文参数或 init 参数

由于内嵌的容器都遵循 Servlet 规范，因此也支持 JSP 页面，并且默认情况下就支持。Spring Boot 可以轻松地更改 JSP 提供者程序，或者甚至完全禁用相关的支持。表 3-7 列出公开的属性。

表 3-7 JSP 相关的服务器属性

属性	说明
server.servlet.jsp.registered	是否注册 JSP servlet，默认为 true
server.servlet.jsp.class-name	JSP servlet 类名称，默认为 org.apache.jasper.servlet.JspServlet，因为 Tomcat 和 Jetty 都使用 Jasper 作为 JSP 实现
server.servlet.jsp.init-parameters	JSP servlet 的上下文参数

■ 注意：
不鼓励并限制将 JSP 与 Spring Boot 应用程序一起使用[11]。

在使用 Spring MVC 时，你可能需要使用 HTTP Session 来存储属性(通常使用 Spring Security 来存储 CSFR 令牌，等)。通用的 servlet 配置还允许你配置 http 会话以及它的存储方式(cookie、URL 等)。相关属性如表 3-8 所示。

表 3-8 HTTP Session 相关的服务器属性

属性	说明
server.servlet.session.timeout	会话超时时间，默认为 30 分钟
server.servlet.session.tracking-modes	会话跟踪模式(可以设置为三种跟踪模式：cookie、url 和 ssl 中的一种或多种)。默认为空，由容器决定
server.servlet.session.persistent	会话数据在重新启动之间是否保持，默认为 false
server.servlet.session.cookie.name	保存会话标识符的 cookie 名称。默认为空，由容器的默认值决定
server.servlet.session.cookie.domain	会话 cookie 的域名称。默认为空，由容器的默认值决定
server.servlet.session.cookie.path	会话 cookie 要使用的路径。默认为空，由容器的默认值决定
server.servlet.session.cookie.comment	会话 cookie 的注释。默认为空，由容器的默认值决定
server.servlet.session.cookie.http-only	会话 cookie 是否只能通过 HTTP 进行访问，默认为空，由容器的默认值决定
server.servlet.session.cookie.secure	Cookie 是否只通过 SSL 发送，默认为空，由容器的默认值决定
server.servlet.session.cookie.max-age	会话 cookie 的生存期。默认为空，由容器的默认值决定
server.servlet.session.session-store-directory.directory	保存持久性 cookie 的目录。必须是现有目录

最后，Spring Boot 通过公开一些属性，使配置 SSL 变得非常容易。关于如何配置 SSL 请参见表 3-9 和 3.8 小节。

表 3-9 SSL 相关的服务器属性

属性	说明
server.ssl.enabled	是否启用 SSL，默认为 true
server.ssl.ciphers	支持的 SSL 密码，默认为空

---

11 https://docs.spring.io/spring-boot/docs/current/reference/html/boot-featuresdeveloping-web-applications.html#boot-features-jsp-limitations.

(续表)

属性	说明
server.ssl.client-auth	是否希望(WANT)或需要(NEED)进行 SSL 客户端身份验证。默认为空
server.ssl.protocol	要使用的 SSL 协议，默认为 TLS
server.ssl.enabled-protocols	启用了哪些 SSL 协议，默认为空
server.ssl.key-alias	用于标识密钥库中的密钥的别名，默认为空
server.ssl.key-password	访问密钥库中的密钥的密码，默认为空
server.ssl.key-store	密钥库的位置，通常是一个 JKS 文件，默认为空
server.ssl.key-store-password	访问密钥存储库的密码，默认为空
server.ssl.key-store-type	密钥库的类型，默认为空
server.ssl.key-store-provider	密钥库的提供商，默认为空
server.ssl.trust-store	信任库的位置
server.ssl.trust-store-password	信任库的访问密码，默认为空
server.ssl.trust-store-type	信任库的类型，默认为空
server.ssl.trust-store-provider	信任库的提供商，默认为空

■ 注意：

表 3-9 中提到的所有属性只有在使用内嵌的容器运行应用程序时才适用。当部署到外部容器(例如，部署一个 WAR 文件)时，这些设置不适用！

因为使用的是 Thymeleaf 库，所以可以禁止注册 JSP servlet，还可以更改端口和压缩功能的配置。在 application.properties 文件中编写以下内容：

```
server.port=8081
server.compression.enabled=true
server.servlet.jsp.registered=false
```

现在当(重新)启动时，应用程序页面(当足够大时)将被压缩，服务器将在端口 8081 而不是 8080 上运行。

■ 注意：

server.servlet.context-path 和 server.servlet.path 之间是有区别的。当查看 URL http://localhost:8080/books 时，它由几个部分组成。第一部分是协议(通常是 http 或 https)，第二部分是需要访问的服务器的地址和端口，第三部分是 context-path(默认部署在根目录下)，第四部分是 servlet-path。server.context-path 是应用程序的主 URL；例如，如果设置 server.context-path=/library，那么可以在 /library URL 上访问整个应用程序(DispatcherServlet 仍然在上下文路径中的根目录/上侦听)。现在，如果设置

server.path=/dispatch，那么就需要使用/library/dispatch/books来访问图书信息。

接下来，如果我们添加另一个 DispatcherServlet，将其路径配置为/services，那么可以通过/library/services 对其进行访问。两个 DispatcherServlets 在/library 的主上下文路径中都是激活的。

2. 更改运行时容器

当包含 spring-boot-starter-web 依赖项时，Spring Boot 将自动包含对 Tomcat 容器的依赖，因为它自己对 spring-boot-starter-tomcat 工件存在依赖关系。为了启用不同的 servlet 容器，需要排除 spring-boot-starter-tomcat，并且需要包含 spring-boot-starter-jetty 或 spring-boot-boot-undertow 两者中的一个。

```
<dependency>
  <groupId>org.springframework.boot</groupId>
  <artifactId>spring-boot-starter-web</artifactId>
  <exclusions>
   <exclusion>
     <groupId>org.springframework.boot</groupId>
     <artifactId>spring-boot-starter-tomcat</artifactId>
  </exclusion>
  </exclusions>
</dependency>
<dependency>
  <groupId>org.springframework.boot</groupId>
  <artifactId>spring-boot-starter-jetty</artifactId>
</dependency>
```

在 Maven 中，可以在<dependency>元素中使用 <exclusion>元素来排除依赖项。

现在，当应用程序启动时，它将启动 Jetty 容器，而不是使用 Tomcat 容器，如图 3-11 所示。

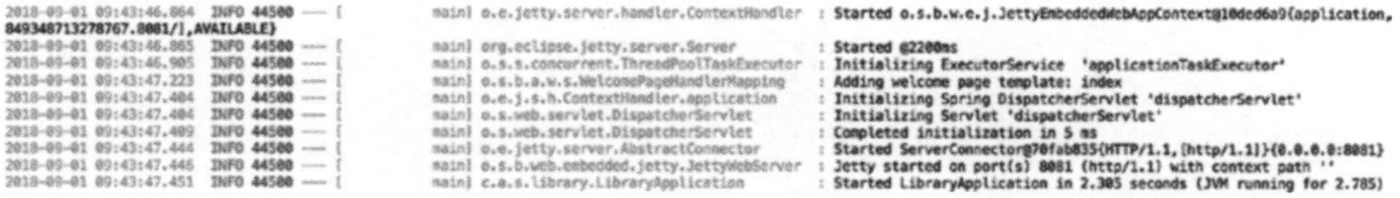

```
2018-09-01 09:43:46.864  INFO 44500 --- [           main] o.e.jetty.server.handler.ContextHandler  : Started o.s.b.w.e.j.JettyEmbeddedWebAppContext@10ded6a9{application,
849348713278767.8081/],AVAILABLE}
2018-09-01 09:43:46.865  INFO 44500 --- [           main] org.eclipse.jetty.server.Server          : Started @2200ms
2018-09-01 09:43:46.905  INFO 44500 --- [           main] o.s.s.concurrent.ThreadPoolTaskExecutor  : Initializing ExecutorService  'applicationTaskExecutor'
2018-09-01 09:43:47.223  INFO 44500 --- [           main] o.s.b.a.w.s.WelcomePageHandlerMapping    : Adding welcome page template: index
2018-09-01 09:43:47.404  INFO 44500 --- [           main] o.e.j.s.h.ContextHandler.application     : Initializing Spring DispatcherServlet 'dispatcherServlet'
2018-09-01 09:43:47.404  INFO 44500 --- [           main] o.s.web.servlet.DispatcherServlet        : Initializing Servlet 'dispatcherServlet'
2018-09-01 09:43:47.409  INFO 44500 --- [           main] o.s.web.servlet.DispatcherServlet        : Completed initialization in 5 ms
2018-09-01 09:43:47.444  INFO 44500 --- [           main] o.e.jetty.server.AbstractConnector       : Started ServerConnector@70fab835{HTTP/1.1,[http/1.1]}{0.0.0.0:8081}
2018-09-01 09:43:47.446  INFO 44500 --- [           main] o.s.b.web.embedded.jetty.JettyWebServer  : Jetty started on port(s) 8081 (http/1.1) with context path ''
2018-09-01 09:43:47.451  INFO 44500 --- [           main] c.a.s.library.LibraryApplication         : Started LibraryApplication in 2.305 seconds (JVM running for 2.785)
```

图 3-11　Jetty 容器的启动日志

3.8 为 Servlet 容器配置 SSL

3.8.1 问题

你希望使用 HTTPS 替代 HTTP 来访问应用程序。

3.8.2 解决方案

获取一个证书，将其放入密钥库中，并使用 server.ssl 名称空间中的属性来配置密钥库。然后，Spring Boot 将自动配置服务器，使其只能通过 HTTPS 访问。

3.8.3 工作原理

使用 server.ssl.key-store 属性(和其他相关属性)，可以将内嵌的容器配置为只接受 HTTPS 连接。在配置 SSL 之前，需要一个证书来保护应用程序。通常，你需要从证书颁发机构，如 VeriSign 或 Let's Encrypt，获得证书。但是，出于演示的目的，你可以使用自签名证书。

1. 创建自签名证书

Java 附带了一个名为 keytool 的工具，它可以完成生成证书等功能。

```
keytool -genkey -keyalg RSA -alias sb2-recipes -keystore
sb2-recipes.pfx-storepass password -validity 3600 -keysize 4096
-storetype pkcs12
```

上述命令将告诉 keytool 工具使用 RSA 算法生成一个密钥，并将其放入名为 sb2-recipes.pfx 的密钥库中，同时使用别名 sb2-recipes 为其命名，该密钥的有效期为 3600 天。当运行这个命令时，keytool 会要求回答几个问题，请根据问题作出相应的回答(或者留空)。之后，keytool 将会生成一个名为 sb2-recipes.pfx 的文件，其中包含证书并受到密码保护。

将这个文件放到 src/main/resources 文件夹中，以便将其打包为应用程序的一部分，同时 Spring Boot 可以轻松地访问它。

■ **警告：**
使用自签名证书将在浏览器中产生一个警告，显示网站不安全且不受到保护，因为证书不是由受信任的权威机构颁发的(参见图 3-12)。

2. 配置 Spring Boot 使用密钥库

Spring Boot 需要了解密钥库，以便能够配置内嵌的容器。为此，需要使用 server.ssl.key-store 属性。还需要指定密钥库的类型(本章使用的密钥库类型是 pkcs12

和密码。

```
server.ssl.key-store=classpath:sb2-recipes.pfx
server.ssl.key-store-type=pkcs12
server.ssl.key-store-password=password
server.ssl.key-password=password
server.ssl.key-alias=sb2-recipes
```

现在，当打开 http://localhost:8080/books.html 页面时，它将通过 https 返回结果(尽管带有一个警告)，如图 3-12 所示。

图 3-12 通过 HTTPS 访问图书馆

3. 同时支持 HTTP 和 HTTPS

默认情况下，Spring Boot 只启动一个连接器：要么是 HTTP 连接器，要么是 HTTPS 连接器，但不能同时启动两个。如果希望同时支持 HTTP 和 HTTPS，则必须手动添加一个附加的连接器。最简单的方法是自己创建 HTTP 连接器并让 Spring Boot 设置 SSL 相关的内容。

首先配置 Spring Boot 在端口 8443 上启动服务器。

```
server.port=8443
```

要向内嵌的 Tomcat 容器添加额外的连接器，需要将 TomcatServletWebServerFactory 类作为 bean 添加到上下文中。通常情况下，Spring Boot 会检测容器并选择要使用的 WebServerFactory 类；但是，由于需要进行自定义，我们需要手动添加它。可以将这个 bean 添加到使用@Configuration 注解的类或 LibraryApplication 类中。

```java
@Bean
public TomcatServletWebServerFactory tomcatServletWebServerFactory(){
  var factory = new TomcatServletWebServerFactory();
  factory.addAdditionalTomcatConnectors(httpConnector());
  return factory;
}
private Connector httpConnector() {
  var connector = new Connector(TomcatServletWebServerFactory
.DEFAULT_PROTOCOL);
  connector.setScheme("http");
  connector.setPort(8080);
  connector.setSecure(false);
  return connector;
}
```

这将在 8080 端口上添加一个额外的连接器。现在可以从 8080 和 8443 两个端口访问该应用程序。使用 Spring Security，可以强制使用 HTTPS(而不是 HTTP)访问应用程序的某些部分。

■ 提示：
如果不想显式地配置 TomcatServletWebServerFactory，也可以使用 BeanPostProcessor 来向 TomcatServletWebServerFactory 注册额外的 Tomcat Connector。通过这种方式，可以为不同的内嵌容器配置两个连接器，而不用绑定到单个容器。

```java
@Bean
public BeanPostProcessor addHttpConnectorProcessor() {
  return new BeanPostProcessor() {
  @Override
  public Object postProcessBeforeInitialization(Object bean, String beanName)
  throws BeansException {
  if (bean instanceof TomcatServletWebServerFactory) {
    var factory = (TomcatServletWebServerFactory) bean;
    factory.addAdditionalTomcatConnectors(httpConnector());
  }
  return bean;
```

    }
  };
}
```

### 4. 重定向 HTTP 到 HTTPS

除了同时支持 HTTP 和 HTTPS 以外，另一种选择是只支持 HTTPS 并将访问请求和响应从 HTTP 重定向到 HTTPS。这种配置与同时支持 HTTP 和 HTTPS 时的配置类似。但是，现在将连接器配置为将所有的访问请求和响应从 8080 端口重定向到 8443 端口。

```java
@Bean
public TomcatServletWebServerFactory tomcatServletWebServerFactory(){
  var factory = new TomcatServletWebServerFactory();
  factory.addAdditionalTomcatConnectors(httpConnector());
  factory.addContextCustomizers(securityCustomizer());
  return factory;
}

private Connector httpConnector() {
  var connector = new Connector(TomcatServletWebServerFactory
.DEFAULT_PROTOCOL);
  connector.setScheme("http");
  connector.setPort(8080);
  connector.setSecure(false);
  connector.setRedirectPort(8443);
  return connector;
}

private TomcatContextCustomizer securityCustomizer() {
  return context -> {
    var securityConstraint = new SecurityConstraint();
    securityConstraint.setUserConstraint("CONFIDENTIAL");
    var collection = new SecurityCollection();
    collection.addPattern("/*");
    securityConstraint.addCollection(collection);
    context.addConstraint(securityConstraint);
  };
}
```
```

上述代码中，httpConnector 方法调用 setRedirectPort 方法设置了重定向端口，这样它就知道该使用哪个端口了。最后，需要使用 SecurityConstraint 来保护所有 URL。通过 Spring Boot，可以使用专用的 TomcatContextCustomizer 类对 Tomcat 的上下文在 Tomcat 启动之前进行后期处理。由于使用了/*模式，这项约束使得一切变成了机密，结果是所有访问请求和响应都会被重定向到 https。此时允许使用 NONE、INTEGRAL 或 CONFIDENTIAL 三种模式之一访问应用程序。

第 4 章

Spring MVC 异步特性

当发布 Servlet API 时，大多数容器的实现方式都是采用每个请求占用一个线程的方式。这意味着在请求处理完成并将响应发送给客户端之前，线程一直处于阻塞状态。

在早期，连接到互联网的设备没有现在那么多。由于设备数量的增加，服务器处理的 HTTP 请求的数量也增加了。由于增加了许多网络应用，处理请求时阻塞线程的处理方式已不再可行。

根据 Servlet 3 规范，容器可以异步处理 HTTP 请求，这是通过释放处理 HTTP 请求的线程实现的。新线程将在后台运行，一旦处理完成，它就会把处理结果写回客户端。这一切都可以在符合 Servlet 3.1 规范的 Servlet 容器上以非阻塞的方式完成。处理过程中，所有需要使用的资源也必须是非阻塞式的。

在过去的几年中还出现了反应式编程的潮流。在 Spring 5 中也可以编写反应式 Web 应用。为了支持编写反应式应用程序，Spring 使用 Project Reactor 实现了 Reactive Streams API。对反应式编程进行全面研究超越了本书的范围。简而言之，反应式编程是一种进行非阻塞函数式编程的方法。

当使用 Web 应用时，服务器会接收请求，然后在服务器上呈现 HTML，并将其发送回客户端。在过去的几年里，呈现 HTML 的工作转移到了客户端。服务器在与客户端通信时，将 JSON、XML 或其他格式的、需要展现的内容返回给客户端。

尽管客户端通过 XmlHttpRequest 对象进行异步调用，但这仍然是一个请求和响应周期。在客户端和服务器之间还有其他通信方式。可以使用服务器发送事件(Server-Sent-Event)进行单向通信。对于全双工通信，可以使用 Web 套接字。

在构建普通的 Web 应用时，需要添加 spring-boot-starter-web 依赖项。当构建反应式应用程序时，需要添加 spring-boot-starter-webflux 依赖项。

4.1 使用控制器和 TaskExecutor 处理异步请求

4.1.1 问题

为了减少 servlet 容器上的吞吐量,你希望异步处理请求。

4.1.2 解决方案

当请求到达服务器时,服务器对其进行同步处理,处理 HTTP 请求的线程会被阻塞。响应保持打开状态,服务器可以向其写入数据。对于异步处理方式,例如,当服务器需要一段时间来处理请求时,服务器可以在后台处理请求,并在完成时向用户返回一个值,这种处理方式相比于阻塞线程的方式来说是非常有用的。

4.1.3 工作原理

Spring MVC 支持方法返回多种类型的数据,表 4-1 中列出了以异步方式返回的数据类型。

表 4-1 异步返回的数据类型

类型	说明
DeferredResult<V>	其他线程返回的异步结果
ListenableFuture<V>	其他线程返回的异步结果,等同于 DeferredResult 类型
CompletionStage<V>/CompletableFuture<V>	其他线程返回的异步结果,等同于 DeferredResult 类型
Callable<V>	计算结束后返回的异步计算结果
ResponseBodyEmitter	用于将多个对象异步写入响应
SseEmitter	用于异步写入服务器发送的事件
StreamingResponseBody	用于向 OutputStream 异步写入

1. 编写异步控制器程序

为了编写异步处理请求的控制器程序,只需要修改其处理请求的方法的返回类型(参见表 4-1)即可。假设调用 HelloWorldController.hello 方法需要等待相当长的时间才能获得返回结果,但是我们不想为此阻塞服务器。

2. 使用 Callable

```
package com.apress.springbootrecipes.library;

import org.springframework.web.bind.annotation.GetMapping;
import org.springframework.web.bind.annotation.RestController;
```

```
import java.util.concurrent.Callable;
import java.util.concurrent.ThreadLocalRandom;

@RestController
public class HelloWorldController {

  @GetMapping
  public Callable<String> hello() {
    return () -> {
      Thread.sleep(ThreadLocalRandom.current().nextInt(5000));
      return "Hello World, from Spring Boot 2!";
    };
  }
}
```

hello 方法现在返回的值是 Callable<String>类型，而不是直接返回 String 类型的值。在新构建的、返回值为 Callable<String>类型的 hello 方法内，代码随机等待一段时间来模拟延迟，然后将消息返回到客户端。

现在，在进行预订的时候，会在日志中看到一些与图 4-1 类似的跟踪信息。

```
2018-09-09 10:06:14.957 DEBUG 82160 --- [nio-8080-exec-1] o.s.web.servlet.DispatcherServlet        : GET "/hello", parameters={}
2018-09-09 10:06:14.972 DEBUG 82160 --- [nio-8080-exec-1] s.w.s.m.m.a.RequestMappingHandlerMapping : Mapped to public java.util.concurrent.Callable<java.lar
2018-09-09 10:06:14.996 DEBUG 82160 --- [nio-8080-exec-1] o.s.w.c.request.async.WebAsyncManager    : Started async request
2018-09-09 10:06:14.998 DEBUG 82160 --- [nio-8080-exec-1] o.s.web.servlet.DispatcherServlet        : Exiting but response remains open for further handling
2018-09-09 10:06:15.708 DEBUG 82160 --- [         task-1] o.s.w.c.request.async.WebAsyncManager    : Async result set, dispatch to /hello
2018-09-09 10:06:15.716 DEBUG 82160 --- [nio-8080-exec-2] o.s.web.servlet.DispatcherServlet        : "ASYNC" dispatch for GET "/hello", parameters={}
2018-09-09 10:06:15.723 DEBUG 82160 --- [nio-8080-exec-2] s.w.s.m.m.a.RequestMappingHandlerAdapter : Resume with async result ["Hello World, from Spring Boc
2018-09-09 10:06:15.730 DEBUG 82160 --- [nio-8080-exec-2] o.s.w.s.m.m.a.RequestResponseBodyMethodProcessor : Using 'text/plain', given [*/*] and supported [text/pla
n]
2018-09-09 10:06:15.731 DEBUG 82160 --- [nio-8080-exec-2] m.m.a.RequestResponseBodyMethodProcessor : Writing ["Hello World, from Spring Boot 2!"]
2018-09-09 10:06:15.740 DEBUG 82160 --- [nio-8080-exec-2] o.s.web.servlet.DispatcherServlet        : Exiting from "ASYNC" dispatch, status 200
```

图 4-1　异步处理的日志输出

在图 4-1 中可以看到，请求处理是在某个线程上启动的(这里是 nio-8080-exec-1 线程)，另有一个线程进行处理并返回结果(这里是 task-1 线程)。最后，请求被再次分发给 DispatcherServlet，以处理另一个线程上的结果(这里是 nio-8080-exec-2 线程)。

3. 使用 CompletableFuture

更改 hello 方法的签名以返回 CompletableFuture<String>类型的值，并使用 TaskExecutor 异步执行代码。

```
@RestController
public class HelloWorldController {

  private final TaskExecutor taskExecutor;
```

```java
    public HelloWorldController(TaskExecutor taskExecutor) {
        this.taskExecutor = taskExecutor;
    }

    @GetMapping
    public CompletableFuture<String> hello() {

    return CompletableFuture.supplyAsync(() -> {
      randomDelay();
      return "Hello World, from Spring Boot 2!";
    }, taskExecutor);
  }

  private void randomDelay() {
    try {
      Thread.sleep(ThreadLocalRandom.current().nextInt(5000)) ;
    } catch (InterruptedException e) {
      Thread.currentThread().interrupt();
    }
  }
}
```

上述代码中，调用 supplyAsync 方法将提交一个任务(或者返回类型为 CompletableFuture<void> 时，调用 runAsync 方法)。它返回一个类型为 CompletableFuture 的值。代码中使用的 supplyAsync 方法同时接收两个参数，Supplier 和 Executor。通过这种方式，可以重用 TaskExecutor 进行异步处理。如果使用 supplyAsync 方法，该方法只接收 Supplier 参数，那么它将从 JVM 上可用的默认 fork/join 池中获取一个 Executor 线程。

当返回 CompletableFuture 类型的值时，可以利用这个数据类型支持的多种特性，例如可以组合和链接多个 CompletableFuture 实例。

4. 测试异步控制器

与常规控制器的测试一样，Spring MVC 测试框架也可以用于测试异步控制器。创建一个测试类并使用 @WebMvcTest(HelloWorldController.class) 和 @RunWith(SpringRunner.class)对其进行注解。@WebMvcTest 将启动一个最小化的 Spring Boot 应用程序，该应用程序包含测试控制器所需的内容。Spring Boot 自动配置 Spring MockMvc 实例，该实例将自动选择相关测试方法进行测试。

```java
@RunWith(SpringRunner.class)
```

```
@WebMvcTest(HelloWorldController.class)
public class HelloWorldControllerTest {

  @Autowired
  private MockMvc mockMvc;

  @Test
  public void testHelloWorldController() throws Exception {
    MvcResult mvcResult = mockMvc.perform(get("/"))
      .andExpect(request().asyncStarted())
      .andDo(MockMvcResultHandlers.log())
      .andReturn();

    mockMvc.perform(asyncDispatch(mvcResult))
      .andExpect(status().isOk())
      .andExpect(content().contentTypeCompatibleWith(TEXT_PLAIN))
      .andExpect(content().string("Hello World,from Spring Boot 2!"));
  }
}
```

异步 Web 测试和常规 Web 测试之间的主要区别在于需要对异步分发进行初始化。首先执行初始请求并验证可以启动异步处理。出于调试的目的，可以使用 MockMvcResultHandlers.log()记录相关日志。接下来，asyncDispatch 开始工作。最后我们可以使用断言判断响应是否和预期的情况一致。

5. 配置异步处理过程

根据需要，你可能还希望为异步请求处理配置一个显式的 TaskExecutor，而不是使用由 Spring Boot 提供的默认 TaskExecutor。

若要配置异步处理，需要重写 WebMvcConfigurer 类的 configureAsyncSupport 方法。重写此方法使你可以访问 AsyncSupportConfigurer 类。这允许你设置(在多个执行器中)要使用的 AsyncTaskExecutor。

```
@SpringBootApplication
public class HelloWorldApplication implements WebMvcConfigurer {

  public static void main(String[] args) {
    SpringApplication.run(HelloWorldApplication.class, args);
  }
```

```java
    @Override
    public void configureAsyncSupport(AsyncSupportConfigurer 
configurer) {
        configurer.setTaskExecutor(mvcTaskExecutor());
    }

    @Bean
    public ThreadPoolTaskExecutor mvcTaskExecutor() {
      ThreadPoolTaskExecutor taskExecutor = new
        ThreadPoolTaskExecutor();
      taskExecutor.setThreadNamePrefix("mvc-task-");
      return taskExecutor;
    }
}
```

4.2 响应回写函数

4.2.1 问题

服务器提供服务或收到多次调用请求，应用程序希望收集这些处理结果并将响应发送给客户端。

4.2.2 解决方案

使用 ResponseBodyEmitter(或它的子类 SseEmitter)收集响应并将其发送给客户端。

4.2.3 工作原理

1. 在一次响应中发送多个结果

Spring MVC 有一个名为 ResponseBodyEmitter 的类，如果你希望在每次响应时向客户端返回多个对象，而不是返回单次调用的结果时，那么这个类非常有用。在发送对象时，使用 HttpMessageConverter 将其转换为一个处理结果。要使用 ResponseBodyEmitter 对象，必须在代码中调用该类的构造函数进行实例化，并通过请求处理方法返回其实例。

使用 orders 方法创建 OrderController 实例，该方法返回 ResponseBodyEmitter 实例并将多个结果逐个发送给客户端。

首先创建 Order 类和 OrderService 类。

```java
public class Order {

  private String id;
  private BigDecimal amount;

  public Order() {
  }

  public Order(String id, BigDecimal amount) {
    this.id=id;
    this.amount = amount;
  }

  public String getId() {
    return id;
  }

  public void setId(String id) {
    this.id = id;
  }

  public BigDecimal getAmount() {
    return amount;
  }

  public void setAmount(BigDecimal amount) {
  this.amount = amount;
  }

  @Override
  public String toString() {
    return String.format("Order [id='%s',amount=%4.2f]",id,amount);
  }
}
```

然后添加一个简单的 OrderService 类来保存订单。

```java
@Service
public class OrderService {
```

```java
  private final List<Order> orders = new ArrayList<>();

  @PostConstruct
  public void setup() {
    createOrders();
  }

  public Iterable<Order> findAll() {
    return List.copyOf(orders);
  }

  private Iterable<Order> createOrders() {
    for (int i = 0; i < 25; i++) {
      this.orders.add(createOrder());
    }
    return orders;
  }

  private Order createOrder() {
    String id = UUID.randomUUID().toString();
    double amount=ThreadLocalRandom.current().nextDouble(1000.00d);
    return new Order(id, BigDecimal.valueOf(amount));
  }
}
```

现在创建 OrderController 类。

```java
@RestController
public class OrderController {

  private final OrderService orderService;

  public OrderController(OrderService orderService) {
    this.orderService = orderService;
  }

  @GetMapping("/orders")
  public ResponseBodyEmitter orders() {
```

```java
      var emitter = new ResponseBodyEmitter();
      var executor = Executors.newSingleThreadExecutor();
      executor.execute(() -> {
        var orders = orderService.findAll();
        try {
          for (var order : orders) {
            randomDelay();
            emitter.send(order);
          }
            emitter.complete();
        } catch (IOException e) {
            emitter.completeWithError(e)
        }
      });
      executor.shutdown();
      return emitter;
    }

    private void randomDelay() {
      try {
        Thread.sleep(ThreadLocalRandom.current().nextInt(150));
      } catch (InterruptedException e) {
        Thread.currentThread().interrupt();
      }
    }
  }
}
```

首先创建了一个 ResponseBodyEmitter 对象，并在方法的最后返回该对象。接下来将执行一个任务，使用 OrderService.findAll 方法查询预定信息。使用 ResponseBodyEmitter 的 send 方法逐个返回此次调用的所有结果。当所有对象都已发送完毕时，需要调用 complete()方法，以便负责发送响应的线程能够完成请求处理，并被释放出来以便处理下一个响应。当出现异常并希望通知用户时，可以调用 completeWithError 方法，该异常将由常规的 Spring MVC 异常处理机制进行处理，此后响应结束。

当使用诸如 httpie 或 curl 之类的工具时，调用 http://localhost:8080/orders 地址时将产生类似于图 4-2 所示的结果。

```
 http :8080/orders -v
GET /orders HTTP/1.1
Accept: */*
Accept-Encoding: gzip, deflate
Connection: keep-alive
Host: localhost:8080
User-Agent: HTTPie/0.9.9

HTTP/1.1 200
Date: Sun, 09 Sep 2018 08:34:38 GMT
Transfer-Encoding: chunked

{"id":"a3e2c659-6d61-4f72-81a2-9386767901c7","amount":964.3303264760648}{"id":"68e353b9-69b0-44cc-a0a5-c9d811a61f5e","amou
nt":630.5819945932226}{"id":"7cd568aa-4168-40cb-996c-051e5fa82457","amount":247.3223213836674}{"id":"964a6124-f993-43eb-ba
17-908bfc5097aa","amount":808.549237068922}{"id":"5ecc62bc-9f3c-473b-9966-d5c9f37555fd","amount":661.3744699952401}{"id":"
f676d1bc-dde0-4f8c-83b5-e3101642c123","amount":857.9818651414307}{"id":"2c4f7232-83f5-41b3-9633-4f71a48e3364","amount":891
.7136746089751}{"id":"d20b0c31-8dcd-46d1-aea3-9d0fac9f14c0","amount":332.4999391686701 5}{"id":"b683f8ee-011b-4390-8973-80a
a17509d7a","amount":800.6509712001965}{"id":"3cd778f0-9e25-4d2a-b29e-92d239a9189d","amount":801.3667308438916}{"id":"abd8e
7da-9d76-42d9-a0f0-0455e0518fb1","amount":554.321571430265}{"id":"7769e9a4-f951-447c-961c-d715c8a18ab8","amount":115.50183
059747442}{"id":"5b1fc94f-853b-4f32-b01c-1f4af7ce41dd","amount":686.6941002940033}{"id":"05d36d52-786a-4a5e-b02a-915958b89
37e","amount":748.0382636058531}{"id":"6564f567-d9ad-4486-b5dd-b77ce1e1b06c","amount":879.4838692645402}{"id":"be6c103d-f0
9a-4afe-9f5f-992d522b42ba","amount":447.45108559657245}{"id":"46a39e54-bb29-43dc-9417-c17fc7462574","amount":419.547649244
9695}{"id":"51a7c85c-b35a-4269-9b52-e2dd7ef2d08c","amount":421.8349910257171}{"id":"35564e73-b7ee-46ca-9438-fdc9637d63d4",
"amount":975.0530454514776}{"id":"6f0eb0fa-1c60-4a9c-8299-1d592204dd30","amount":338.3559152812614}{"id":"5542e1a6-d500-44
4d-b889-ad0cd98d17a7","amount":402.66337491090167}{"id":"47087519-3e98-4f10-8719-61aa581f06a6","amount":702.9394273793404}
{"id":"11b96b15-d4eb-4f56-b14b-1b3f9ff77d61","amount":176.0339216118647}{"id":"697b74c3-8e35-4514-beb9-3e35d4a7b935","amou
nt":297.7516547391046}{"id":"c058b00d-b09f-442f-8f7b-3c8469f49720","amount":588.2887029367946}
~
```

图 4-2　ResponseBodyEmitter 的输出结果

最后，为这个控制器编写一个测试。用@RunWith(SpringRunner.class)和@WebMvcTest (OrderController.class)注解一个类以便用其进行 Web 切片测试。OrderController 类需要一个 OrderServic 实例，通过在 OrderService 字段上使用@MockBean 注解获取模拟实例。

```
@RunWith(SpringRunner.class)
@WebMvcTest(OrderController.class)
public class OrderControllerTest {

  @Autowired
  private MockMvc mockMvc;

  @MockBean
  private OrderService orderService;

  @Test
  public void foo() throws Exception {
    when(orderService.findAll())
      .thenReturn(List.of(new Order("1234", BigDecimal.TEN)));
    MvcResult mvcResult =
      mockMvc.perform(get("/orders"))
        .andExpect(request().asyncStarted())
        .andDo(MockMvcResultHandlers.log())
```

```
        .andReturn();

    mockMvc.perform(asyncDispatch(mvcResult))
        .andDo(MockMvcResultHandlers.log())
        .andExpect(status().isOk())
        .andExpect(content().json("{\"id\":\"1234\",\"amount\":10}"));
    }
}
```

测试方法首先在模拟的 OrderService 对象上注册行为,以返回 Order 对象的单个实例。接下来,使用 MockMvc 在/orders 端点上执行 get 操作。由于这是一个异步控制器,应用程序将开始以异步的方式处理请求。然后,模拟异步分发并用断言判断实际的响应。结果应该是一个 JSON 元素,包含 id 和 amount 两个字段。

2. 将多个结果当作事件发送

SseEmitter 可以将事件从服务器发送到客户端。服务器发送事件是指从服务器发送到客户端的消息。这类消息的 Content-Type 报头为 text/event-stream。事件消息非常短小,只有 4 个字段,如表 4-2 所示。

表 4-2 服务器发送事件中允许出现的字段

字段	描述
id	事件的 ID
event	事件的类型
data	事件数据
retry	事件流重启连接的时间

为了通过请求处理方法发送事件,需要创建一个 SseEmitter 实例并从请求处理方法返回该实例。然后使用该实例的 send 方法将各个事件发送到客户端。

```
@GetMapping("/orders")
public SseEmitter orders() {
  SseEmitter emitter = new SseEmitter();
  ExecutorService executor = Executors.newSingleThreadExecutor();
  executor.execute(() -> {
    var orders = orderService.findAll();
    try {
      for (var order : orders) {
        randomDelay();
        emitter.send(order);
      }
```

```
            emitter.complete();
        } catch (IOException e) {
            emitter.completeWithError(e);
        }
    });
    executor.shutdown();
    return emitter;
}
```

■ 注意：
在将每个事件发送到客户端时都会有延迟。上述代码这样做仅用于演示传送多个事件，但在实际代码中不会这样操作。

现在，当访问 http://localhost:8080/orders 时，可以看到事件一个接一个地出现，如图 4-3 所示。

```
▶ http :8080/orders
HTTP/1.1 200
Content-Type: text/event-stream;charset=UTF-8
Date: Wed, 11 Jul 2018 10:06:56 GMT
Transfer-Encoding: chunked

data:{"id":"941a263f-3f48-49cb-9801-ca931623a355","amount":730.4383307235909}

data:{"id":"0b0aefda-afdf-47da-853d-25ce67977597","amount":154.7487508307738}

data:{"id":"03504d98-ce57-4950-9627-8dd582b0f43f","amount":689.9288545237567}

data:{"id":"bc2da6aa-d6dc-4201-b880-35f114ba6f51","amount":581.8911724684582}

data:{"id":"a25f2d92-ff06-4b00-b08b-f9b85fa4355c","amount":779.151699482505}

data:{"id":"a70c3311-cc88-4a82-93ab-06e25a490cbb","amount":198.71754307506419}

data:{"id":"334bc6c5-8cc5-40f7-bc3b-1035b99dcc68","amount":371.8884523273436}

data:{"id":"416ddcad-4149-42ec-bf79-50a8f10c81a9","amount":955.9374666581973}

data:{"id":"e64d821f-6a6f-4c81-9698-99e135cc8fd4","amount":549.3973604937846}

data:{"id":"a220b6a9-53f0-4a3e-b34c-938966007727","amount":179.1484242467024}

data:{"id":"598448a5-2120-4f37-9896-b66f1a09d0d5","amount":138.47217193639062}

data:{"id":"82224f61-34f4-4ee4-8066-91e9f119c36c","amount":976.5754411843043}
```

图 4-3　服务器发送事件的演示结果

注意，Content-Type 头的值为 text/event-stream，表示我们得到了一个事件流。这个流可以保持打开状态并接收事件通知。编写的每个对象都通过 HttpMessage-Converter 转换为 JSON 对象。每个对象都作为事件数据写入 data 字段中。

如果希望向事件添加更多信息（即填写表 4-2 中提到的其他字段之一），需要使用 SseEventBuilder 对象。SseEmitter 类的 event()工厂方法可以创建 SseEventBuilder 实例，通过该实例填充 id 和 event 字段。

```java
@GetMapping("/orders")
public SseEmitter orders() {
  SseEmitter emitter = new SseEmitter();
  ExecutorService executor = Executors.newSingleThreadExecutor();
  executor.execute(() -> {
    var orders = orderService.findAll();
    try {
      for (var order : orders) {
        randomDelay();
        var eventBuilder = event();
        emitter.send(
          eventBuilder
            .data(order)
            .name("order-created")
            .id(String.valueOf(order.hashCode())));
      }
      emitter.complete();
    } catch (IOException e) {
      emitter.completeWithError(e);
    }
  });
  executor.shutdown();
  return emitter;
}
```

现在，当访问 http://localhost:8080/orders 时，服务器发送回来的事件包含了 id、event 和 data 三个字段。见图 4-4。

■ 注意：

微软的浏览器(Internet Explorer或Edge)不支持服务器发送事件，要在Microsoft 浏览器中使用事件流，必须使用polyfill插件[1]。

---

1 https://github.com/remy/polyfills.

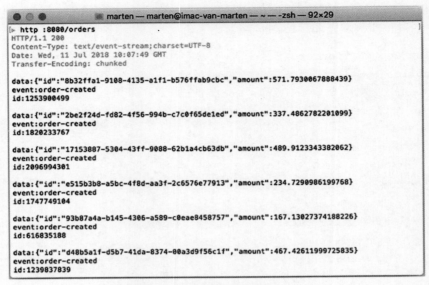

图 4-4　终端输出的订单数据

4.3　WebSocket

4.3.1　问题

你希望通过网络从客户端到服务器进行双向通信。

4.3.2　解决方案

使用 WebSocket 在客户端和服务器之间进行通信。与 HTTP 不同，WebSocket 提供了全双工通信。

4.3.3　工作原理

对 WebSocket 进行详细的介绍超出了本书的范围，然而，值得一提的是，HTTP 和 WebSocket 之间的关系实际上并不大。WebSocket 唯一用到 HTTP 的地方是最初的握手使用 HTTP，将连接从普通的 HTTP 升级到一个 TCP 套接字连接。

1. 配置 Spring Boot 支持 WebSocket

配置的第一步是添加 spring-boot-starter-websocket 依赖项，这个依赖项将注入所需的依赖，并将在 Spring Boot 中自动配置 WebSocket 支持。

为了启用 WebSocket，只需要将@EnableWebSocket 注解添加到应用程序中。需要将该注解添加到@SpringBootApplication 注解类或@Configuration 注解类上，如以下代码所示。

```
@SpringBootApplication
@EnableWebSocket
public class EchoApplication implements WebSocketConfigurer { ... }
```

2. 创建 WebSocketHandler

为了处理 WebSocket 消息和生命周期事件(即握手，建立连接等)，需要创建一个 WebSocketHandler 对象并将其注册到某个端点 URL。

如果希望直接实现 WebSocketHandler 接口的话，需要实现该接口定义的 5 个方法，如表 4-3 所示。然而，Spring 已经提供了很好的类层次结构，你可以根据自己的情况决定选择使用哪个类。在编写自定义处理程序时，扩展 TextWebSocketHandler 或 BinaryWebSocketHandler 中的一个就足够了；顾名思义，前者可以处理文本消息，后者可以处理二进制消息。

表 4-3　WebSocketHandler 定义的 5 个方法

方法	说明
afterConnectionEstablished	在 WebSocket 连接已经打开且可以使用时触发
handleMessage	当处理器对应的 WebSocket 消息到达时触发
handleTransportError	发生错误时触发
afterConnectionClosed	WebSocket 连接关闭后触发
supportsPartialMessages	处理器是否支持分段消息：如果设置为 true，一个 WebSocket 消息可以通过多次调用 handleMessage 进行接收

如下代码通过扩展 TextWebSocketHandler 类来创建 EchoHandler 类，并实现 afterConnectionEstablished 方法和 handleTextMessage 方法。

```
package com.apress.springbootrecipes.echo;

import org.springframework.web.socket.TextMessage;
import org.springframework.web.socket.WebSocketSession;
import org.springframework.web.socket.handler.TextWebSocketHandler;

public class EchoHandler extends TextWebSocketHandler {
@Override
  public void afterConnectionEstablished(WebSocketSession session)
  throws Exception {
    session.sendMessage(new TextMessage("CONNECTION ESTABLISHED"));
  }
```

```java
@Override
protected void handleTextMessage(WebSocketSession session,
                    TextMessage message) throws Exception {
    var msg = message.getPayload();
    session.sendMessage(new TextMessage("RECEIVED: " + msg));
}
}
```

建立连接后，服务器会向客户端发送一个 TextMessage，内容是"CONNECTION ESTABLISHED"，告知客户端连接已经建立。当服务器接收到 TextMessage 时，将提取有效负载(实际的消息)并以 RECEIVED 作为前缀将其发送回客户端。

接下来，需要在一个 URI 上注册这个处理器，为此，可以创建一个用@Configuration 注解的类来实现 WebSocketConfigurer 接口并在 registerWebSocketHandlers 方法中进行注册。接下来将此接口添加到 EchoApplication 类。

```java
package com.apress.springbootrecipes.echo;

import org.springframework.boot.SpringApplication;
import org.springframework.boot.autoconfigure.SpringBootApplication;
import org.springframework.context.annotation.Bean;
import org.springframework.web.socket.config.annotation.EnableWebSocket;
import org.springframework.web.socket.config.annotation.WebSocketConfigurer;
import org.springframework.web.socket.config.annotation.WebSocketHandlerRegistry;

@SpringBootApplication
@EnableWebSocket
public class EchoApplication implements WebSocketConfigurer {
 public static void main(String[] args) {
    SpringApplication.run(EchoApplication.class, args);
 }

 @Bean
 public EchoHandler echoHandler() {
   return new EchoHandler();
 }
```

```java
@Override
public void registerWebSocketHandlers(WebSocketHandlerRegistry
registry) {
    registry.addHandler(echoHandler(), "/echo");
}
}
```

首先将 EchoHandler 注册为 bean，以便可以将其注册到一个 URI。在 registerWebSocketHandlers 方法中，可以使用 WebSocketHandlerRegistry 来注册处理器。使用 addHandler 方法，可以将处理器注册到一个 URI，在本例中是/echo。有了这个配置，可以使用 URL ws://localhost:8080/echo 从客户端打开一个 WebSocket 连接。

现在服务器已经就绪，我们需要一个客户端来连接到 WebSocket 端点。为此，需要编写一些 JavaScript 和 HTML 代码。编写如下代码所示的 app.js 文件并将其放到 src/main/resources/static 目录中。

```javascript
var ws = null;
var url = "ws://localhost:8080/echo";

function setConnected(connected) {
    document.getElementById('connect').disabled = connected;
    document.getElementById('disconnect').disabled = !connected;
    document.getElementById('echo').disabled = !connected;
}

function connect() {
    ws = new WebSocket(url);

    ws.onopen = function () {
        setConnected(true);
    };
  ws.onmessage = function (event) {
        log(event.data);
    };

    ws.onclose = function (event) {
        setConnected(false);
        log('Info: Closing Connection.');
```

```
    };
}

function disconnect() {
    if (ws != null) {
        ws.close();
        ws = null;
    }
    setConnected(false);
}
function echo() {
    if (ws != null) {
        var message = document.getElementById('message').value;
        log('Sent: ' + message);
        ws.send(message);
    } else {
        alert('connection not established, please connect.');
    }
}

function log(message) {
    var console = document.getElementById('logging');
    var p = document.createElement('p');
    p.appendChild(document.createTextNode(message));
    console.appendChild(p);
    while (console.childNodes.length > 12) {
        console.removeChild(console.firstChild);
    }
    console.scrollTop = console.scrollHeight;
}
```

上述代码中有多个函数。第一个函数是 connect，按下 index.html 页面(稍后说明)上的 Connect 按钮时将触发该函数，这将打开到 ws://localhost:8080/echo 的 WebSocket 连接，服务器侧触发的处理器是前面创建和注册的 EchoHandler。连接到服务器将创建一个 WebSocket JavaScript 对象，这将使你能够侦听客户端上的消息。这里定义了 onopen、onmessage 和 onclose 3 个回调函数。最重要的是 onmessage，因为只要消息从服务器传入，就会调用 onmessage 函数；该函数只调用 log 函数，log 函数将接收到的消息添加到 logging 元素中并输出到屏幕上。

接下来是 disconnect 函数，该函数将关闭 WebSocket 连接并清理 JavaScript 对象。最后还有一个 echo 函数，每当按下 Echo Message 按钮时，就会调用这个函数。此时，代码中指定的消息会发送到服务器(并最终返回到客户端)。

为了使用 app.js，在 src/main/resources/static 目录中添加以下 index.html 文件。

```html
<!DOCTYPE html>
<html>
<head>
    <link type="text/css" rel="stylesheet"href="https://cdnjs.cloudflare.com/ajax/libs/semantic-ui/2.2.10/semantic.min.css" />
    <script type="text/javascript" src="app.js"></script>
</head>
<body>
<div>
    <div id="connect-container" class="ui centered grid">
      <div class="row">
          <button id="connect" onclick="connect();" class="ui green button ">Connect</button>
          <button id="disconnect" disabled="disabled" onclick="disconnect();" class="ui red button">Disconnect</button>
      </div>
      <div class="row">
           <textarea id="message" style="width: 350px" class="ui input" placeholder="Message to Echo"></textarea>
      </div>
    <div class="row">
          <button id="echo" onclick="echo();" disabled="disabled" class="ui button">Echo message</button>
     </div>
   </div>
    <div id="console-container">
        <h3>Logging</h3>
        <div id="logging"></div>
   </div>
</div>
</body>
</html>
```

现在，部署应用程序后，可以通过访问 http://localhost:8080 连接到上述 echo WebSocket 服务，发送一条消息到服务器并接收服务器的返回消息，如图 4-5 所示。

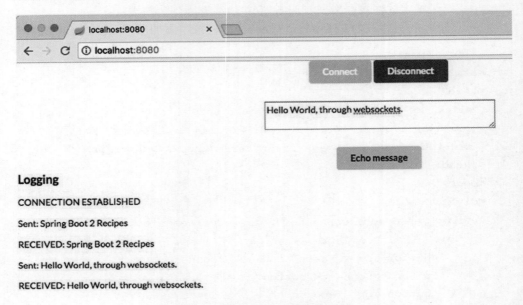

图 4-5　WebSocket 客户端的输出

3. WebSocketHandler 单元测试

现在希望编写一个单元测试，以确保 EchoHandler 实际执行你希望它做的事情。为了对 EchoHandler 进行单位测试，编写如下测试代码并使用 Mockito(或其他模拟框架)来模拟 WebSocket 部分。

```
package com.apress.springbootrecipes.echo;

import org.junit.Test;
import org.springframework.web.socket.TextMessage;
import org.springframework.web.socket.WebSocketSession;

import static org.mockito.ArgumentMatchers.eq;
import static org.mockito.Mockito.*;

public class EchoHandlerTest {

    private final EchoHandler handler = new EchoHandler();
```

```
@Test
public void shouldEchoMessage() throws Exception {
    var mockSession = mock(WebSocketSession.class);
    var msg = new TextMessage("Hello World!");
    handler.handleTextMessage(mockSession, msg);

    verify(mockSession, times(1))
      .sendMessage(eq(new TextMessage("RECEIVED: Hello World!")));
}
```

该测试创建了一个 EchoHandler 实例,并在测试方法中模拟了 WebSocketSession 实例。TextMessage 的构造非常简单。使用 TextMessage 和模拟的 WebSocketSession 实例调用 handleTextMessage 方法。为了验证服务器侧代码按照我们的要求执行了相关功能,测试最后使用 verify 函数确认模拟的 WebSocketSession 实例的返回结果。

4. 集成测试 WebSocket

编写集成测试要复杂一些。需要与服务器建立一个 WebSocket 连接,手动发送消息并检查响应。但在此之前,需要先启动服务器。

要启动服务器,可以使用 @RunWith(SpringRunner.class) 和 @SpringBootTest (WebEnvironment = SpringBootTest.WebEnvironment.RANDOM_PORT)注解一个类。我们需要实际启动应用程序,而不是启动默认的 WebEnvironment.MOCK 模拟环境。在 int 字段上使用@LocalServerPort 注解来获取实际的端口号。构造要连接的 URI 时需要这个端口号。

```
@RunWith(SpringRunner.class)
@SpringBootTest(webEnvironment = SpringBootTest.WebEnvironment
.RANDOM_PORT)
public class EchoHandlerIntegrationTest {

    @LocalServerPort
    private int port;
}
```

使用默认的 Java WebSocket API 测试 WebSocket 相当容易。要进行测试,可以编写一个基本的 WebSocket 客户端来记录接收到的消息和会话。用@ClientEndpoint 注解一个类,并用@OnOpen、@OnClose 和@OnMessage 来注解类中相应的方法。当连接被打开或关闭,或接收到消息时,带注解的相应方法将分别被回调。接下来有两个帮助方法:sendTextAndWait 和 closeAndWait。sendTextAndWait 方法将使用 Session 对象发送消息并等待响应。closeAndWait 方法将关闭会话并等待确认。最后,还有一些 getter 方法用于接收状态并在测试方法中对返回结果进行验证。

在 EchoHandlerIntegrationTest 类中添加一个内嵌的静态类 SimpleTestClient-Endpoint，如以下代码所示。

```java
@ClientEndpoint
public static class SimpleTestClientEndpoint {

    private List<String> received = new ArrayList<>();
    private Session session;
    private CloseReason closeReason;
    private boolean closed = false;

    @OnOpen
    public void onOpen(Session session) {
        this.session = session;
    }

    @OnClose
    public void onClose(Session session, CloseReason reason) {
        this.closeReason = reason;
        this.closed = true;
    }

    @OnMessage
    public void onMessage(String message) {
        this.received.add(message);
    }

    public void sendTextAndWait(String text, long timeout)
            throws IOException, InterruptedException {
        var current = received.size();
        session.getBasicRemote().sendText(text);
        wait(() -> received.size() == current, timeout);
    }

    public void closeAndWait(long timeout)
            throws IOException, InterruptedException {
        if (session != null && !closed) {
            session.close();
```

```
    }
    wait(() -> closeReason == null, timeout);
  }

  private void wait(Supplier<Boolean> condition, long timeout)
          throws InterruptedException {
    var waited = 0;
    while (condition.get() && waited < timeout) {
      Thread.sleep(1);
      waited += 1;
    }
  }

  public CloseReason getCloseReason() {
      return closeReason;
  }

  public List<String> getReceived() {
    return this.received;
  }

  public boolean isClosed() {
      return closed;
  }
}
```

现在相关的辅助类已经就绪，可以编写测试了。首先，使用 WebSocket ContainerProvider 获取容器。在建立实际连接之前，必须使用 port 字段构造 URI。接下来，将 SimpleTestClientEndpoint 类的实例 testClient 作为参数调用 connectToServer 方法来连接到服务器。连接后，客户端向服务器发送一条文本消息并等待一段时间，然后关闭连接(即释放资源)。最后一步是使用断言对接收到的消息进行判断。在这个测试中，我们希望正好收到两条消息。

```
@Test
public void sendAndReceiveMessage() throws Exception {
  var container = ContainerProvider.getWebSocketContainer();
  var uri = URI.create("ws://localhost:" + port + "/echo");
  var testClient = new SimpleTestClientEndpoint();
  container.connectToServer(testClient, uri);
```

```
    testClient.sendTextAndWait("Hello World!", 200);
    testClient.closeAndWait(2);

    assertThat(testClient.getReceived())
   .containsExactly("CONNECTION ESTABLISHED", "RECEIVED: Hello
    World!");

}
```

4.4 在 WebSocket 上使用 STOMP

4.4.1 问题

你希望在 WebSocket 上使用 STOMP(Simple/Streaming Text Oriented Message Protocol，简单/流式面向文本的消息协议)与服务器进行通信。

4.4.2 解决方案

配置消息代理，在由@Controller 注解的类中，使用@MessageMapping 注解处理消息的方法。

4.4.3 工作原理

让我们讨论一下如何使用 WebSocket 创建一个或多或少包含消息传递的应用程序。尽管可以按原样使用 WebSocket 协议，但该协议允许使用子协议。其中一个由 Spring WebSocket 支持子协议的是 STOMP。

STOMP 是一个非常简单的面向文本的协议。它是为像 Ruby 和 Python 这样的脚本语言创建的，用于连接到消息代理。STOMP 可以在任何可靠的双向网络协议上使用，如 TCP 和 WebSocket。该协议本身是面向文本的，但是消息的有效负载并没有严格地限制这一点，它还可以传输二进制数据。

在通过 Spring WebSocket 的支持配置和使用 STOMP 时，WebSocket 应用程序充当所有已连接客户端的代理。代理可以是内存中的代理，也可以是一个真正成熟的企业解决方案，该解决方案支持 STOMP 协议(如 RabbitMQ 或 ActiveMQ)。在后一种情况下，Spring WebSocket 应用程序充当实际代理的中继。要在 WebSocket 上添加消息传递功能，Spring 使用 Spring Messaging 抽象。

1. 使用 STOMP 和消息映射

为了能够接收消息，需要在 @Controller 注解的类中使用@MessageMapping 标记一个方法，并告诉它将从哪个目的地接收消息。让我们修改 EchoHandler(本章前面提供的示例)以使用该注解，如以下代码所示。

```
package com.apress.springbootrecipes.echo;

import org.springframework.messaging.handler.annotation.MessageMapping;
import org.springframework.messaging.handler.annotation.SendTo;
import org.springframework.stereotype.Controller;

@Controller
public class EchoHandler {

    @MessageMapping("/echo")
    @SendTo("/topic/echo")
    public String echo(String msg) {
        return "RECEIVED: " + msg;
    }
}
```

当在 /app/echo 目的地上接收到消息时，它将被传递到@MessageMapping 注解的方法。另外，请注意方法上的@SendTo("/topic/echo")，该注解指示 Spring 将结果，一个字符串，放在前面的主题上。

配置消息代理并添加用于接收消息的端点。为此，将@EnableWebSocketMessageBroker 注解添加到 EchoApplication 类上，并让它实现 WebSocketMessageBrokerConfigurer 接口。

```
package com.apress.springbootrecipes.echo;

import org.springframework.boot.SpringApplication;
import org.springframework.boot.autoconfigure.SpringBootApplication;
import org.springframework.messaging.simp.config.MessageBrokerRegistry;
import org.springframework.web.socket.config.annotation.EnableWebSocketMessageBroker;
import org.springframework.web.socket.config.annotation.StompEndpointRegistry;
import org.springframework.web.socket.config.annotation.WebSocketMessageBrokerConfigurer;

@SpringBootApplication
```

```
@EnableWebSocketMessageBroker
public class EchoApplication implements
WebSocketMessageBrokerConfigurer {

 @Override
 public void configureMessageBroker(MessageBrokerRegistry registry) {
    registry.enableSimpleBroker("/topic");
    registry.setApplicationDestinationPrefixes("/app");
 }

 @Override
 public void registerStompEndpoints(StompEndpointRegistry registry) {
    registry.addEndpoint("/echo-endpoint");
 }

 public static void main(String[] args) {
     SpringApplication.run(EchoApplication.class, args);
 }
}
```

@EnableWebSocketMessageBroker 注解将启用通过 WebSocket 传递消息的功能，代理在 configureMessageBroker 方法中进行配置。这里使用了简单的消息代理，若要连接到企业级代理，需要使用 registry.enableStompBrokerRelay 方法连接到实际的代理。

为了区分由代理或应用程序处理的消息，这里使用了不同的前缀，如图 4-6 所示。目标地址中以 /topic 开头的任何内容都将被传递给代理，以 /app 开头的目标地址上的任何内容都将被发送到消息处理器(即@MessageMapping 注解的方法)。

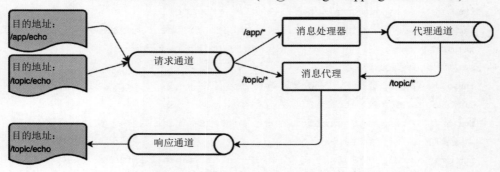

图 4-6　WebSocket STOMP 消息通道和路由

第 4 章　Spring MVC 异步特性

上述代码的最后一部分注册一个 WebSocket 端点，该端点侦听即将到达的 STOMP 消息。在本例中，该端点被映射到 /echo-endpoint。

下面修改客户端以使用 STOMP 而不是直接使用 WebSocket。HTML 代码几乎可以保持不变。需要再添加一个库才能在浏览器中使用 STOMP，本书使用 webstomp-client 库[2]，但你可以使用不同的库。

```html
<head>
  <link type="text/css" rel="stylesheet" href="https://cdnjs.cloudflare.com/ajax/libs/semantic-ui/2.2.10/semantic.min.css" />

  <script type="text/javascript" src="webstomp.js"></script>
  <script type="text/javascript" src="app.js"> </script>
</head>
```

app.js 文件的变化最大，如以下代码所示。

```javascript
var ws = null;
var url = "ws://localhost:8080/echo";

function setConnected(connected) {
    document.getElementById('connect').disabled = connected;
    document.getElementById('disconnect').disabled = !connected;
    document.getElementById('echo').disabled = !connected;
}

function connect() {
  ws = webstomp.client(url, {protocols: ['v11.stomp', 'v12.stomp']});
  ws.connect({}, function(frame) {
      setConnected(true);
      log(frame);
      ws.subscribe('/topic/echo', function(message){
          log(message.body);
      })
  });
}

  function disconnect() {
```

---

2 https://github.com/JSteunou/Webstomp-client.

```
    if (ws != null) {
        ws.disconnect();
        ws = null;
    }
    setConnected(false);
}
function echo() {
    if (ws != null) {
        var message = document.getElementById('message').value;
        log('Sent: ' + message);
        ws.send("/app/echo", message);
    } else {
        alert('connection not established, please connect.');
    }
}

function log(message) {
    var console = document.getElementById('logging');
    var p = document.createElement('p');
    p.appendChild(document.createTextNode(message));
    console.appendChild(p);
    while (console.childNodes.length > 12) {
        console.removeChild(console.firstChild);
    }
    console.scrollTop = console.scrollHeight;
}
```

connect 函数现在使用 webstomp.client 函数来创建一个 STOMP 客户端到代理的连接。当连接建立之后，客户端将订阅/topic/echo 并接收放置在该主题上的消息。echo 函数已经修改为使用客户端的 send 方法将消息发送到目标地址/app/echo。

在启动应用程序并打开客户端时，你仍然能够发送和接收消息，但是现在使用的是 STOMP 子协议。你甚至可以连接多个浏览器，每个浏览器都会看到 /topic/echo 目标地址上的消息，因为它是一个主题类型的消息发布模型。

在编写带@MessageMapping 注解的方法时，可以使用不同的方法参数和注解(如表 4-4 所示)来接收关于消息的更多或更少的信息。默认情况下，只有一个参数会映射到消息的有效载荷；MessageConverter 用于将消息的有效载荷转换为所需的类型。

表 4-4　@MessageMapping 注解的方法支持的方法参数和注解

类型	说明
Message	包含消息头和消息体的实际底层消息
@Payload	消息的有效载荷(默认情况)；参数也可以用 @Validated 进行注解以进行验证
@Header	从 Message 中获取指定的消息头
@Headers	可用于注解 Map 参数以获取所有消息头
MessageHeaders	Message 的所有消息头
Principal	当前用户(如果已设置)

2. 为处理器编写单元测试

为处理器编写单元测试相对简单。处理器的 echo 方法接受一个 String 类型的参数并返回 String 类型的值。要进行测试，只需要调用该方法并使用断言验证返回的值。

```
package com.apress.springbootrecipes.echo;

import org.junit.Test;

import static org.assertj.core.api.Assertions.assertThat;

public class EchoHandlerTest {

  private final EchoHandler handler = new EchoHandler();

  @Test
  public void shouldEchoMessage() throws Exception {

    var msg = "Hello World!";
    assertThat(handler.echo(msg)).isEqualTo("RECEIVED: " + msg);
  }
}
```

3. 使用 STOMP 进行集成测试

要进行集成测试，需要启动应用程序，可能需要在一个随机端口上构造一个 STOMP 客户端，并发送一条消息。测试的最后一部分(发送和接收消息)是异步的，因此进行测试会有一点困难。首先，在随机端口上启动应用程序，并在测试类中获取该随机端口。

```
@RunWith(SpringRunner.class)
@SpringBootTest(webEnvironment=SpringBootTest.WebEnvironment.
  RANDOM_PORT)
public class EchoHandlerIntegrationTest {

  @LocalServerPort
  private int port;
}
```

上述代码中，@RunWith 指示 JUnit 使用 SpringRunner 来执行测试。由于将 webEnvironment 设置为 SpringBootTest.WebEnvironment.RANDOM_PORT，@SpringBootTest 注解指示 Spring Boot 在一个随机端口上启动该应用程序。在 int 字段上添加@LocalServerPort 注解，我们可以得到实际的端口。

接下来，为了测试可以使用 Spring 客户端库中的 WebSocketStompClient 连接到服务器并订阅一个主题。要接收关于某个主题的消息，需要编写一个 StompSessionHandlerAdapter 的扩展类，并重写 getPayloadType 和 handleFrame 方法。由于我们并不打算重用这个类，因此可以将其声明为 EchoHandlerIntegrationTest 类中的一个静态嵌套类。

```
private static class TestStompFrameHandler extends
StompSessionHandlerAdapter {

  private final CompletableFuture<String> answer;

  private TestStompFrameHandler(CompletableFuture<String> answer) {
    this.answer = answer;
  }

  @Override
  public Type getPayloadType(StompHeaders headers) {
    return byte[].class;
  }

  @Override
  public void handleFrame(StompHeaders headers, Object payload) {
    answer.complete(new String((byte[]) payload));
  }
}
```

由于 WebSocket 的异步特性，我们需要一些手段来阻塞线程，直到收到响应。为了传递结果，我们使用了 CompletableFuture 类型的参数。在使用断言时，我们可以使用 get 方法来阻塞线程，直到收到一个响应(或会话超时)。

对于本次测试，我们需要创建一个 STOMP 客户端并使用该客户端来连接到 STOMP 代理，订阅 /topic/echo 以接收消息，最后发送消息并使用断言检查结果。

```java
@RunWith(SpringRunner.class)
@SpringBootTest(webEnvironment = SpringBootTest.WebEnvironment.RANDOM_PORT)
public class EchoHandlerIntegrationTest {

    @LocalServerPort
    private int port;

    private WebSocketStompClient stompClient;

    private List<StompSession> sessions = new ArrayList<>();

    @Before
    public void setup() {
        var webSocketClient = new StandardWebSocketClient();
        stompClient = new WebSocketStompClient(webSocketClient);
    }

    @After
    public void cleanUp() {
        this.sessions.forEach(StompSession::disconnect);
        this.sessions.clear();
    }
    @Test
    public void shouldSendAndReceiveMessage() throws Exception {
        CompletableFuture<String> answer = new CompletableFuture<>();
        var stompSession = connectAndSubscribe(answer);

        stompSession.send("/app/echo", "Hello World!".getBytes());

        var result = answer.get(1, TimeUnit.SECONDS);
        assertThat(result).isEqualTo("RECEIVED: Hello World!");
    }
```

```java
    private StompSession connectAndSubscribe(CompletableFuture
<String>answer)
        throws InterruptedException,ExecutionException,TimeoutException {
    var uri = "ws://localhost:" + port + "/echo-endpoint";
    var stompSession =
      stompClient.connect(uri, new StompSessionHandlerAdapter() {})
        .get(1, TimeUnit.SECONDS);

    stompSession.subscribe("/topic/echo",
        new TestStompFrameHandler(answer));
    this.sessions.add(stompSession);
    return stompSession;
    }
      ...
}
```

如上述代码所示，在@Before 注解的方法中创建 WebSocketStompClient 实例，该实例使用 StandardWebSocketClient 实例作为传输工具。在@After 注解的方法中，我们清除了所有与已连接到代理的会话(只是为了确保不占用任何资源)。connectAndSubscribe 方法使用 WebSocketStompClient 连接到代理，在这里它被配置为等待 1 秒后连接。然后，使用 StompSession 订阅前面创建的 TestStompFrame-Handler，订阅时使用了 connectAndSubscribe 方法接收的 CompletableFuture 类型的参数。当然，测试中不使用辅助方法也可以，但不使用这些辅助方法通常需要更多的集成测试，同时使用辅助方法便于创建可重用的代码片段。最后的实际测试步骤是：STOMP 客户端首先连接并订阅主题，然后将消息"Hello World!"发送到服务器并期望在 1 秒(或更短的时间)内得到一个响应。最后，按照预期的结果检查接收到的结果。

第 5 章

Spring WebFlux 特性

5.1 使用 Spring WebFlux 开发反应式应用

5.1.1 问题

你希望使用 Spring WebFlux 开发一个简单的反应式 Web 应用，以了解该框架的基本概念和配置。

5.1.2 解决方案

Spring WebFlux 最底层的组件是 HttpHandler，它是一个只有 handle 方法的接口。

```
public interface HttpHandler {

  Mono<Void> handle(ServerHttpRequest request, ServerHttpResponse
    response);
}
```

handle 方法返回一个 Mono <Void>类型的值，表示以反应式的方式返回 void 类型的值。此函数使用来自 org.springframework.http.server.reactive 包的 ServerHttpRequest 和 ServerHttpResponse 作为参数。这两个参数也是接口，实际调用时将根据运行应用程序所使用的容器来创建接口的实例。为此，容器通常提供多个可选的适配器。当应用程序在支持非阻塞式 IO 的 Servlet 3.1 容器上运行时，Spring 使用 ServletHttpHandlerAdapter 类(或其子类之一)将普通的 Servlet 环境适配成反应式环境。当应用程序在像 Netty[1] 这样原生的反应式引擎上运行时，使用 ReactorHttpHandlerAdapter 进行适配。

当 Web 请求发送到 Spring WebFlux 应用程序时，HandlerAdapter 首先接收请求。

---

[1] https://netty.io。

然后它组织在 Spring 的应用程序上下文中配置的不同组件,这些组件都是处理请求所需的。

要在 Spring WebFlux 中定义控制器类,必须使用 @Controller 或@RestController 注解标记类。这与 Spring MVC 中的做法相同(参见第 3 章)。

当带@Controller 注解的类(即控制器类)接收到请求时,它会寻找合适的处理器方法来处理该请求。这就要求控制器类通过一个或多个处理器映射将每个请求映射到合适的处理器方法。为此,使用@RequestMapping 对控制器类的方法进行注解,从而将相应的方法标记为处理器方法。

这些处理器方法的签名是开放的,这与标准类的签名相同。只要符合相关的命名规则,可以为处理器方法指定任意名称,并定义各种方法参数。同样,处理器方法可以返回多种类型的值(例如字符串或空值),这取决于它实现的应用程序逻辑。以下是部分有效参数类型的列表,以便你对处理器方法可返回的数据类型有个初步的认识。

(1) ServerHttpRequest 或 ServerHttpResponse 对象;
(2) 使用@RequestParam 标记的 URL 中任意类型的请求参数;
(3) 使用@CookieValue 标记的、传入请求中包含的 cookie 值;
(4) 使用@RequestHeader 标记的任意类型的请求消息报头;
(5) 使用@RequestAttribute 标记的任意类型的请求属性;
(6) Map 或 ModelMap 类型,处理器方法使用这两种类型的对象向模型中添加属性;
(7) 会话中的 WebSession。

一旦控制器类选择了适当的处理器方法,它就会调用处理器方法的逻辑来处理请求。通常,控制器的逻辑会调用后端服务来处理请求。此外,处理器方法的逻辑可能会在大量的输入参数(如 ServerHttpRequest、Map 或 Errors)中添加或删除部分内容,形成新的消息,然后根据逻辑传递给其他服务或方法进行处理。

在处理器方法处理完请求之后,它将控制委托给一个视图,该视图充当处理器方法的返回值。为了灵活起见,处理器方法的返回值并不表示视图的最终呈现(例如,user.html 或 report.pdf),而是表示逻辑视图(例如 user 或 report)——注意没有文件扩展名。

处理器方法的返回值可以是一个表示逻辑视图名称的 String 对象,也可以是 void 类型。当返回值为 void 类型时,根据处理器方法或控制器的名称确定默认的逻辑视图名称。

将信息从控制器传递到视图与处理器方法返回逻辑视图的名称(String 或 void)并不相关,因为视图可以访问处理器方法的输入参数。

例如,如果处理器方法将 Map 和 Model 对象作为输入参数并在内部逻辑中修改了这两个对象,处理器方法返回的视图仍然可以访问修改前的 Map 和 Model 对象。

当控制器类接收到一个视图时,它通过视图解析器将逻辑视图名称解析为特定的视图实现(例如,user.html 或 report.fmt)。视图解析器是一个在 Web 应用上下文中配置的 bean，它实现了 ViewResolver 接口。视图解析器的职责是为逻辑视图名称返回一个特定的视图实现。

一旦控制器类将视图名称解析为视图实现，则根据视图实现的设计，它将呈现由控制器的处理器方法返回的对象(例如 ServerHttpRequest、Map、Errors 或 WebSession)。视图的职责是将在处理器方法的逻辑中添加的对象展现给用户。

5.1.3 工作原理

让我们为 4.1 小节的 HelloWorldApplication 编写一个反应式版本。

```
package com.apress.springbootrecipes.helloworld;

import org.springframework.web.bind.annotation.GetMapping;
import org.springframework.web.bind.annotation.RestController;
import reactor.core.publisher.Mono;

@RestController
public class HelloWorldController {
 @GetMapping
  public Mono<String> hello() {
     return Mono.just("Hello World, from Spring Boot 2!");
  }
}
```

注意 hello 方法的返回类型是 Mono<String>，而不是普通的 String。Mono 类型是编写反应式应用程序的关键。

1. 配置 Spring WebFlux 应用程序

为了能够以反应式的方式处理请求，需要启用 webflux。这是通过添加 spring-boot-starter-webflux 依赖项完成的。

```
<dependency>
  <groupId>org.springframework.boot</groupId>
  <artifactId>spring-boot-starter-webflux</artifactId>
</dependency>
```

此依赖项包含了所需的相关依赖，例如 spring-webflux 和 Reactor 项目 (http://www.project Reactor.io)依赖项。它还包括一个反应式运行时，默认情况下是 Netty。

现在一切都已配置好，最后要做的事情就是创建一个应用程序类。

```
package com.apress.springbootrecipes.library;
```

```
import org.springframework.boot.SpringApplication;
import org.springframework.boot.autoconfigure
.SpringBootApplication;

@SpringBootApplication
public class HelloWorldApplication {

  public static void main(String[] args) {
      SpringApplication.run(HelloWorldApplication.class, args);
  }
}
```

Spring Boot 将检测到反应式运行时，并使用以 server 为前缀的属性(即 server.*)对其进行配置。

2. 创建 Spring WebFlux 控制器

任何未实现特定接口或未扩展特定基类的类都可以作为控制器类，只需要使用@Controller 或@RestController 对其进行注解即可。控制器中可以定义一个或多个处理器方法来处理单个或多个操作。处理器方法的签名应足够灵活，以便接收不同的参数。

```
package com.apress.springbootrecipes.helloworld;


import org.springframework.web.bind.annotation.GetMapping;
import org.springframework.web.bind.annotation.RestController;
import reactor.core.publisher.Mono;

@RestController
public class HelloWorldController {

    @GetMapping("/hello")
    public Mono<String> hello() {
        return Mono.just("Hello World, from Reactive Spring Boot 2!");
    }
}
```

注解@GetMapping 用于将 hello 方法标记为控制器的 HTTP GET 处理器方法。值得一提的是，如果没有声明默认的 HTTP GET 处理器方法，应用程序就会抛出

ServletException 异常，因此控制器至少需要包含一个 URL 路由和一个处理器方法。由于@GetMapping 中的表达式，该方法绑定到/hello。

当在/hello 上发起 GET 请求时，它将以反应式的方式返回 "Hello World, from Reactive Spring Boot 2!" 消息，尽管客户端不会注意到这一点。对于客户端来说，它仍然是一个常规的 HTTP 请求。

3. 反应式控制器的单元测试

有三种方法可以对控制器进行集成测试。第一种方法是简单地编写一个测试，其中创建 HelloWorldController 实例，调用方法，并使用断言对结果进行验证。第二种方法是使用@WebFluxTest 注解来创建测试。后者将启动包含网络基础设施的最小应用程序上下文，可以使用 MockMvc 来测试控制器。最后一种方法介于简单单元测试和全面集成测试之间。

```
package com.apress.springbootrecipes.helloworld;

import org.junit.Test;
import reactor.core.publisher.Mono;
import reactor.test.StepVerifier;

public class HelloWorldControllerTest {

  private final HelloWorldController controller = new HelloWorldController();

  @Test
  public void shouldSayHello() {
     Mono<String> result = controller.hello();

     StepVerifier.create(result)
        .expectNext("Hello World, from Reactive Spring Boot 2!")
        .verifyComplete();
  }
}
```

上述代码完成一个基本的单元测试。代码实例化控制器，并简单地调用要测试的方法。它使用 reactive-test 模块的 StepVerifier 方法使测试更容易。先调用 hello 方法，然后使用 StepVerifier 验证结果。

通过如下代码添加 reactive-test 依赖项：

```
<dependency>
```

```xml
<groupId>io.projectreactor</groupId>
<artifactId>reactor-test</artifactId>
<scope>test</scope>
</dependency>
```

第二种方法是使用@WebFluxTest 对特定的控制器进行注解。

```java
@RunWith(SpringRunner.class)
@WebFluxTest(HelloWorldController.class)
public class HelloWorldControllerSliceTest {

    @Autowired
    private WebTestClient webClient;

    @Test
    public void shouldSayHello() {

        webClient.get().uri("/hello").accept(MediaType.TEXT_PLAIN)
                .exchange()
                  .expectStatus().isOk()
                  .expectBody(String.class)
                  .isEqualTo("Hello World,from Reactive Spring Boot 2!");
    }
}
```

这个测试将启动一个最小化的 Spring Boot 上下文，并自动检测项目中所有与 Web 相关的 bean，如使用@ControllerAdvice、@Controller 注解的类等。不需要直接调用 HelloWorldController，该测试使用特殊的 WebTestClient 来声明一个请求(即 webClient.get().uri("/hello"))，并使用 exchange()非阻塞式地发送消息。最后，对响应进行验证。该请求的响应状态应该是 OK，并包含指定的消息主体。

4. 反应式控制器的集成测试

集成测试的代码与上一节中使用@WebFluxTest 进行的测试看起来非常类似。主要的区别是使用了@SpringBootTest 注解，而不是@WebFluxText 注解。使用 @SpringBootTest 将启动整个应用程序，包括所有其他 bean(服务、存储库等)。通过 webEnvironment 属性可以指定要使用的环境，该属性的值可以设置为 RANDOM_PORT、MOCK(默认值)、DEFINED_PORT 或 NONE。在这里，我们使用了一个随机端口，并再次使用 WebTestClient 来触发请求。

```java
@RunWith(SpringRunner.class)
```

```
@SpringBootTest(webEnvironment = SpringBootTest.WebEnvironment
.RANDOM_PORT)
public class HelloWorldControllerIntegrationTest {

    @Autowired
    private WebTestClient webClient;

    @Test
    public void shouldSayHello() {

        webClient.get().uri("/hello").accept(MediaType.TEXT_PLAIN)
            .exchange()
                .expectStatus().isOk()
                .expectBody(String.class)
                .isEqualTo("Hello World, from Reactive Spring Boot 2!");
    }
}
```

请求被发送到内嵌的服务器，然后验证结果具有正确的状态码和消息主体。

■ **注意：**
当将webEnvironment属性设置为MOCK(也是默认值)时，必须添加@Auto-ConfigureWebTestClient注解以便能通过WebTestClient进行测试。

5.2 发布和使用反应式 Rest 服务

5.2.1 问题

你需要编写一个反应式的 REST 端点，该端点生成 JSON 对象。

5.2.2 解决方案

与常规的@RestController 注解类一样，反应式服务可以返回一个常规对象或对象列表，这些对象将被发送到客户端。为了使服务变成反应式的，必须使用相应的反应式数据类型，即 Mono 或 Flux，来封装这些返回值。

5.2.3 工作原理

下面编写一个反应式的 OrderService。每个对外开放的方法要么返回

Mono<Order>类型的值,要么返回 Flux<Order>类型的值。

```
package com.apress.springbootrecipes.order;

@Service
public class OrderService {

  private final Map<String,Order> orders=new ConcurrentHashMap<>(10);

  @PostConstruct
  public void init() {
    OrderGenerator generator = new OrderGenerator();
    for (int i = 0; i < 25; i++ ) {
      var order = generator.generate();
      orders.put(order.getId(), order);
    }
  }

  public Mono<Order> findById(String id) {
    return Mono.justOrEmpty(orders.get(id));
  }

  public Mono<Order> save(Mono<Order> order) {
    return order.map(this::save);
  }

  private Order save(Order order) {
    orders.put(order.getId(), order);
    return order;
  }

  public Flux<Order> orders() {
    return Flux.fromIterable(orders.values()).delayElements
    (Duration.ofMillis(128));
  }
}
```

OrderService 在启动时创建 25 个随机订单。它使用一些简单的方法来获得或保存订单。当检索所有订单时,将延迟 128 毫秒。接下来,创建基本的 Order 类和

OrderGenerator 类。

```java
package com.apress.springbootrecipes.order;
// Imports omitted
public class Order {

  private String id;
  private BigDecimal amount;

  public Order() {
  }

  public Order(String id, BigDecimal amount) {
      this.id=id;
      this.amount = amount;
  }

  public String getId() {
      return id;
  }

  public void setId(String id) {
      this.id = id;
  }

  public BigDecimal getAmount() {
      return amount;
  }

  public void setAmount(BigDecimal amount) {
      this.amount = amount;
  }

  @Override
  public String toString() {
      return String.format("Order[id='%s',amount=%4.2f]",id,amount);
  }

  @Override
```

```java
    public boolean equals(Object o) {
        if (this == o) return true;
        if (o == null || getClass() != o.getClass()) return false;
        Order order = (Order) o;
        return Objects.equals(id, order.id) &&
                Objects.equals(amount, order.amount);
    }

    @Override
    public int hashCode() {
        return Objects.hash(id, amount);
    }
}
```

上述代码是完整的 Order 类，它只包含 id 和 amount 两个属性。OrderGenerator 是用于创建 Order 实例的简单组件，代码如下所示。

```java
package com.apress.springbootrecipes.order;

import java.math.BigDecimal;
import java.util.UUID;
import java.util.concurrent.ThreadLocalRandom;

public class OrderGenerator {

  public Order generate() {
      var amount = ThreadLocalRandom.current().nextDouble(1000.00);
      return new Order(UUID.randomUUID().toString(),
                      BigDecimal.valueOf(amount));
  }
}
```

为了将 Order 发布为 REST 资源，需要创建一个 OrderController 类，如下所示。

```java
package com.apress.springbootrecipes.order.web;

import com.apress.springbootrecipes.order.Order;
import com.apress.springbootrecipes.order.OrderService;
import org.springframework.web.bind.annotation.*;
import reactor.core.publisher.Flux;
```

```java
import reactor.core.publisher.Mono;

@RestController
@RequestMapping("/orders")
public class OrderController {

  private final OrderService orderService;

  OrderController(OrderService orderService) {
      this.orderService = orderService;
  }

  @PostMapping
  public Mono<Order> store(@RequestBody Mono<Order> order) {
      return orderService.save(order);
  }

  @GetMapping("/{id}")
  public Mono<Order> find(@PathVariable("id") String id) {
      return orderService.findById(id);
  }

  @GetMapping
  public Flux<Order> list() {
      return orderService.orders();
  }
}
```

OrderController 映射到 /orders 目标地址并支持列出所有订单或单个订单，以及添加/修改订单。为了启动上述几个类，还需要编写一个简单的 OrderApplication 类作为程序入口。

```java
package com.apress.springbootrecipes.order;

import org.springframework.boot.SpringApplication;
import org.springframework.boot.autoconfigure.SpringBootApplication;

@SpringBootApplication
```

```
public class OrderApplication {
 public static void main(String[] args) {
      SpringApplication.run(OrderApplication.class, args);
  }
}
```

使用诸如 curl 或 httpie 之类的工具可以对端点进行查询。

命令 http http://localhost:8080/orders--stream 应该列出系统中的所有订单(如图 5-1 所示)，命令 http http://localhost:8080/orders/{some-id}应该列出一个订单。

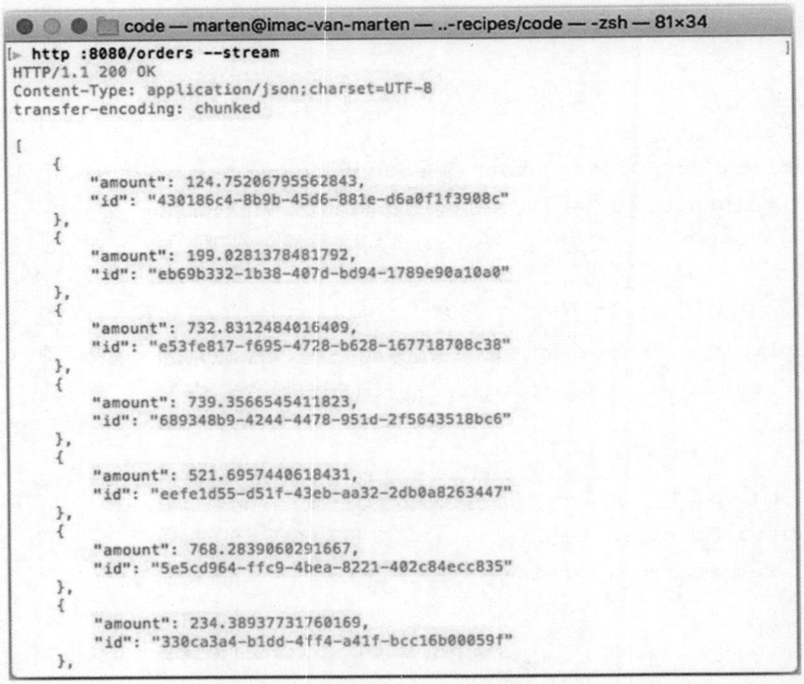

```
code — marten@imac-van-marten — ..-recipes/code — -zsh — 81×34
▶ http :8080/orders --stream
HTTP/1.1 200 OK
Content-Type: application/json;charset=UTF-8
transfer-encoding: chunked

[
    {
        "amount": 124.75206795562843,
        "id": "430186c4-8b9b-45d6-881e-d6a0f1f3908c"
    },
    {
        "amount": 199.0281378481792,
        "id": "eb69b332-1b38-407d-bd94-1789e90a10a0"
    },
    {
        "amount": 732.8312484016409,
        "id": "e53fe817-f695-4728-b628-167718708c38"
    },
    {
        "amount": 739.3566545411823,
        "id": "689348b9-4244-4478-951d-2f5643518bc6"
    },
    {
        "amount": 521.6957440618431,
        "id": "eefe1d55-d51f-43eb-aa32-2db0a8263447"
    },
    {
        "amount": 768.2839060291667,
        "id": "5e5cd964-ffc9-4bea-8221-402c84ecc835"
    },
    {
        "amount": 234.38937731760169,
        "id": "330ca3a4-b1dd-4ff4-a41f-bcc16b00059f"
    },
```

图 5-1　获取所有订单的结果

1. JSON 流

查询/orders 获得的结果不是以流的形式返回的，而是阻塞式的访问。服务首先收集所有的结果，然后再将它们发送出去。如果查看图 5-1 中的 Content-Type 头，可以看到它被设置为 application/json。为了以流的形式返回结果，应该将 Content-Type 头设置为 application/stream+json。为此，需要修改控制器的 list 方法，如下所示。

```
@GetMapping(produces = MediaType.APPLICATION_STREAM_JSON_VALUE)
public Flux<Order> list() {
```

```
    return orderService.orders();
}
```

注意代码中的 produces=MediaType.APPLICATION_STREAM_JSON_VALUE。这指示 Spring 在部分结果准备好时即开始流化传输。重新启动应用程序，然后再次发出命令 http http://localhost:8080/orders--stream。现在，结果将以流的形式逐渐地传入，直到没有更多的订单信息可接收，如图 5-2 所示。

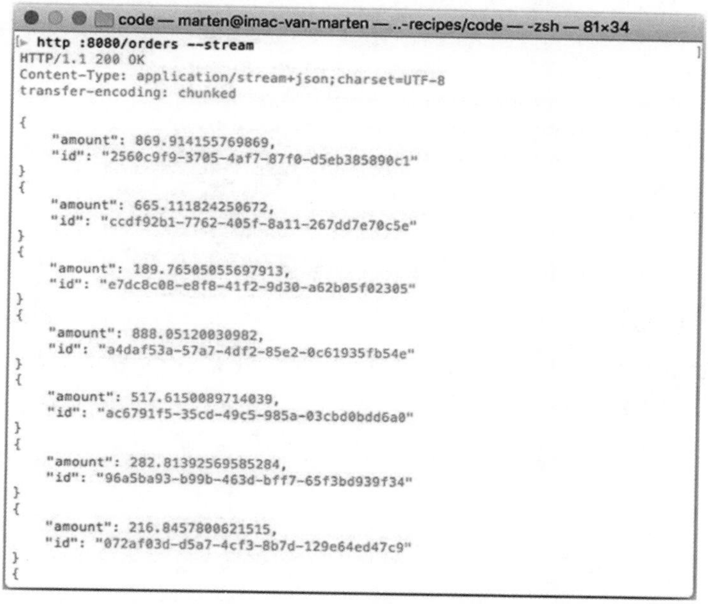

图 5-2　获取所有订单的结果

结果也稍有变化：现在返回的是单个订单(参见图 5-2)，而不是返回一组订单(参见图 5-1)。

2. 服务器发送的事件

除了使用 JSON 流以外，还可以使用服务器发送事件。在 WebFlux 中将消息发送方式切换为服务器发送事件非常简单，只需要将 @GetMapping 注解的方法的 produces 属性设置为 MediaType.TEXT_EVENT_STREAM_VALUE 即可。完成上述修改后，服务器将会以事件的形式发送消息，如以下代码所示。

```
@GetMapping(produces = MediaType.TEXT_EVENT_STREAM_VALUE)
public Flux<Order> list() {
    return orderService.orders();
}
```

重新启动应用程序，然后再次发出命令 http http://localhost:8080/orders --stream。

Spring Boot 2 攻略

结果将以事件的形式流化写入，直到没有更多的订单信息可接收，如图 5-3 所示。请注意，Content-Type 头已经变为 text/event-stream。

```
▶ http --stream :8080/orders
HTTP/1.1 200 OK
Content-Type: text/event-stream;charset=UTF-8
transfer-encoding: chunked

data:{"id":"c2d6772d-882d-4675-a651-4570a9af0125","amount":297.2396366549731}
data:{"id":"fddf6cd7-dbe9-4ff0-8f6f-dfb8ee3663df","amount":305.2656834693431}
data:{"id":"2f4a4c14-2ecc-4c18-b551-9bfb87167052","amount":696.706842452895}
data:{"id":"94c1772e-42f7-4de9-bd5a-af5a5e59df62","amount":0.1775470518441402}
data:{"id":"136fece3-538d-4336-978d-bd890dcabcf3","amount":93.55790629640903}
data:{"id":"91bb34ab-021f-4507-ae20-9a16a947b561","amount":924.8909365656588}
data:{"id":"ebeb3aad-0df1-4877-bd83-8cbf04cee72e","amount":103.94281039673325}
data:{"id":"e4002f19-2215-454f-b451-88637f886079","amount":194.99013045150792}
data:{"id":"55f42c4d-832b-4fef-b2dd-54fc6179b0bc","amount":920.3041222593154}
data:{"id":"a6503643-defa-4c2c-acfe-de6be42e09e9","amount":121.79381700959424}
```

图 5-3 以事件的形式流化输出所有订单

3. 编写集成测试

使用 WebTestClient 和 MOCK Web 环境为 OrderController 编写集成测试非常容易，也可以使用 RANDOM_PORT 代替 MOCK。

```
package com.apress.springbootrecipes.order.web;

import com.apress.springbootrecipes.order.Order;
import org.junit.Test;
import org.junit.runner.RunWith;
import org.springframework.beans.factory.annotation.Autowired;
import org.springframework.boot.test.autoconfigure.web.reactive.AutoConfigureWebTestClient;
import org.springframework.boot.test.context.SpringBootTest;
import org.springframework.test.annotation.DirtiesContext;
import org.springframework.test.context.junit4.SpringRunner;
import org.springframework.test.web.reactive.server.WebTestClient;

import java.math.BigDecimal;

@RunWith(SpringRunner.class)
@SpringBootTest(webEnvironment = SpringBootTest
  .WebEnvironment.MOCK)
@AutoConfigureWebTestClient
```

```java
@DirtiesContext(classMode=DirtiesContext.ClassMode.AFTER_EACH_TEST_METHOD)
public class OrderControllerIntegrationTest {

    @Autowired
    private WebTestClient webTestClient;

    @Test
    public void listOrders() {

        webTestClient.get().uri("/orders")
                .exchange()
                  .expectStatus().isOk()
                  .expectBodyList(Order.class).hasSize(25);
    }

    @Test
    public void addAndGetOrder() {
    var order = new Order("test1", BigDecimal.valueOf(1234.56));
    webTestClient.post().uri("/orders").syncBody(order)
            .exchange()
              .expectStatus().isOk()
              .expectBody(Order.class).isEqualTo(order);

    webTestClient.get().uri("/orders/{id}", order.getId())
            .exchange()
            .expectStatus().isOk()
            .expectBody(Order.class).isEqualTo(order);
    }
}
```

上述代码中添加了@DirtiesContext 注解，因为 OrderService 是一个有状态的 bean。因此在向集合添加一个 Order 之后，需要为下一次测试重置它。因为在模拟环境中使用了@SpringBootTest 注解，测试将启动完整的应用程序。为了获取 WebTestClient 实例，在模拟环境中需要使用@AutoConfigureWebTestClient 注解。使用 WebTestClient 构建请求并将其发送到服务器很容易。在接收到响应之后，代码对结果进行验证。

5.3 使用 Thymeleaf 作为模板引擎

5.3.1 问题

你希望在基于 WebFlux 的应用程序中呈现视图。

5.3.2 解决方案

使用 Thymeleaf 创建视图，以反应式的方式返回视图名称并填充模型。

5.3.3 工作原理

为应用程序添加 spring-boot-starter-webflux 和 spring-boot-starter-thymeleaf 依赖项。这两个依赖项将让 Spring Boot 自动配置 Thymeleaf，以便在 WebFlux 应用程序中使用该模板。

```
<dependency>
    <groupId>org.springframework.boot</groupId>
    <artifactId>spring-boot-starter-webflux</artifactId>
</dependency>
<dependency>
    <groupId>org.springframework.boot</groupId>
    <artifactId>spring-boot-starter-thymeleaf</artifactId>
</dependency>
```

在增加上述依赖项之后，Spring Boot 导入 Thymeleaf 库和 Thymeleaf Spring Dialect 库，以便两个库可以很好地进行集成。由于这两个库的存在，Spring Boot 将自动配置 ThymeleafReactiveViewResolver。

ThymeleafReactiveViewResolver 需要 Thymeleaf 库的 IspringWebFluxTemplate-Engine 类以便解析和呈现视图。Spring Boot 将使用 SpringDialect 预先配置好一个特殊的 SpringWebFluxTemplateEngine，这样就可以在 Thymeleaf 页面中使用 SpEL 语法了。

为了配置 Thymeleaf，Spring Boot 在 Spring.Thymeleaf 和 Spring.Thymeleaf.reactive 名称空间中公开表 5-1 所示的几个属性。

表 5-1 通用的 Thymeleaf 属性

属性	说明
spring.thymeleaf.prefix	ViewResolver 使用的前缀，默认值为 classpath:/templates/
spring.thymeleaf.suffix	ViewResolver 使用的后缀，默认值为 .html

(续表)

属性	说明
spring.thymeleaf.encoding	模板的编码，默认为 UTF-8
spring.thymeleaf.check-template	在展现之前检查模板是否存在，默认为 true
spring.thymeleaf.check-template-location	检查保存模板的路径是否存在，默认为 true
spring.thymeleaf.mode	要使用的 Thymeleaf TemplateMode，默认为 HTML
spring.thymeleaf.cache	是否缓存已解析的模板，默认为 true
spring.thymeleaf.template-resolver-order	ViewResolver 的优先级，默认为 1
spring.thymeleaf.view-names	ViewResolver 可以解析的视图名称(多个名称之间用逗号分隔)
spring.thymeleaf.excludeview-names	无法解析的视图名称(多个名称之间用逗号分隔)
spring.thymeleaf.enabled	是否启用 Thymeleaf 模板，默认为 true
spring.thymeleaf.enable-spring-el-compiler	启用 SpEL 表达式的编译，默认为 false
spring.thymeleaf.reactive.max-chunk-size	用于写入响应的数据缓冲区的最大空间，以字节为单位。默认为 0
spring.thymeleaf.reactive.media-types	视图技术支持的媒体类型，例如 text/html
spring.thymeleaf.reactive.full-mode-view-names	逗号分隔的视图名称列表，这些视图名称应以 full 模式操作，默认为空。full 模式基本上就是阻塞模式
spring.thymeleaf.reactive.chunked-mode-view-names	应以分块模式操作的视图名称列表，多个视图名称之间以逗号分隔

1. 使用 Thymeleaf 视图

首先，在 src/main/resources/templates 目录下创建 index.html 页面，代码如下所示。

```
<!DOCTYPE html>
<html xmlns:th="http://www.thymeleaf.org">
<head>
    <meta charset="UTF-8">
    <title>Spring Boot - Orders</title>
</head>
<body>

<h1>Order Management System</h1>

<a th:href="@{/orders}" href="#">List of orders</a>

</body>
</html>
```

这个页面将呈现一个简单的页面，其中只有一个链接，指向/orders 目标地址。这个 URL 是使用 th:href 标记呈现的，它将把/orders 扩展为一个适当的 URL。接下来，编写一个控制器来选择要呈现的页面。

```
package com.apress.springbootrecipes.order.web

@Controller
class IndexController {

  @GetMapping
  public String index() {
      return "index";
  }
}
```

编写 OrderController 类，它会返回 orders/list 作为视图的名称，并将 Flux<Order> 添加到模型中。

```
package com.apress.springbootrecipes.order.web

@Controller
@RequestMapping("/orders")
class OrderController {

  private final OrderService orderService;

  OrderController(OrderService orderService) {
      this.orderService = orderService;
  }

  @GetMapping
  public Mono<String> list(Model model) {
      var orders = orderService.orders();
      model.addAttribute("orders", orders);
      return Mono.just("orders/list");
  }
}
```

Order 类、OrderService 类以及相关对象的代码请查看 5.2 小节。

现在控制器和其他需要的组件都准备好了，还需要一个视图。在 src/main/resources/templates/orders 目录中创建 list.html 文件。此页面将呈现一个表格，其中显示不同订单的 id 和数量。

```html
<!DOCTYPE html>
<html xmlns:th="http://www.thymeleaf.org">
<head>
  <meta charset="UTF-8">
  <title>Orders</title>
</head>
<body>
<h1>Orders</h1>

  <table>
    <thead>
      <tr>
        <th></th>
        <th>Id</th>
        <th>Amount</th>
      </tr>
    </thead>
    <tbody>
      <tr th:each="order : ${orders}">
        <td th:text="${orderStat.count}">1</td>
        <td th:text="${order.id}"></td>
        <td th:text="${#numbers.formatCurrency(order.amount)}"
            style="text-align: right"></td>
      </tr>
    </tbody>
  </table>
</body>
</html>
```

最后，编写启动应用程序的 OrderApplication 类。

```
@SpringBootApplication
public class OrderApplication {

  public static void main(String[] args) {
      SpringApplication.run(OrderApplication.class, args);
```

}
}

现在，当启动应用程序并单击链接时，将看到一个显示订单的页面，如图 5-4 所示。

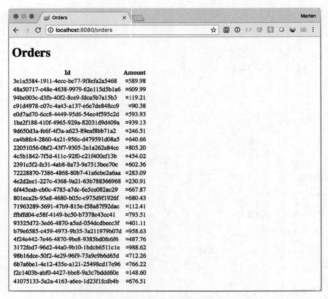

图 5-4 订单列表

2. 增强应用程序的反应速度

在运行应用程序并打开页面时，你必须等待一段时间，然后才能看到页面。当呈现一个页面时，如果模型包含了 Flux，默认情况下，客户端会一直等待直到应用程序处理完整个 Flux。之后，客户端才开始呈现页面。客户端的行为基本上就像你检索的是一个 Collection，而不是 Flux。为了更快地开始展现页面并获得页面流，需要将 Flux 封装在一个 ReactiveDataDriverContextVariable 对象中。

```
@GetMapping
public Mono<String> list(Model model) {
  var orders = orderService.orders();
  model.addAttribute("orders",new ReactiveDataDriverContext-
    Variable(orders,5));
  return Mono.just("orders/list");
}
```

list 方法看起来几乎没变化，但是请注意，Flux 被包装在 ReactiveDataDriver-ContextVariable 对象中。它将在收到 5 个元素后立即开始呈现，并将继续呈现页面，直到收到所有内容。现在，当打开订单页面时，你将看到表格不断变长，直到所有订单都已展现完。

5.4 WebFlux 和 WebSocket

5.4.1 问题

你希望在一个反应式的应用程序中使用 WebSocket。

5.4.2 解决方案

添加依赖项 javax.websocket-api，并使用 Reactive WebSocketHandler 接口实现该处理器程序。

5.4.3 工作原理

在 spring-boot-starter-webflux 之后添加依赖项 javax.websocket-api 就足以在一个反应式应用程序中支持运行 WebSocket。

```xml
<dependency>
  <groupId>org.springframework.boot</groupId>
  <artifactId>spring-boot-starter-webflux</artifactId>
</dependency>
<dependency>
  <groupId>javax.websocket</groupId>
  <artifactId>javax.websocket-api</artifactId>
  <version>1.1</version>
</dependency>
```

Spring WebFlux 支持 WebSocket 1.1 版本。如果 WebSocket 是 1.0 版本，代码不会运行。

接下来，使用 WebSocketHandler 实现一个反应式处理器程序。

```java
package com.apress.springbootrecipes.echo;

import org.springframework.web.reactive.socket.WebSocketHandler;
import org.springframework.web.reactive.socket.WebSocketSession;
import reactor.core.publisher.Mono;

public class EchoHandler implements WebSocketHandler {
```

```java
@Override
public Mono<Void> handle(WebSocketSession session) {
  return session.send(
          session.receive()
              .map( msg -> "RECEIVED: " + msg.getPayloadAsText())
              .map(session::textMessage));
}
}
```

EchoHandler 将接收一条消息，为消息添加前缀"RECEIVED："之后返回消息(对于非反应式版本，请参见 4.4 小节)。

使用 WebFlux 设置 WebSocket 需要注册一些组件。首先我们需要处理器，该处理器程序需要使用 SimpleUrlHandlerMapping 映射到一个 URL。我们需要能调用处理器程序的对象，这里使用的是 WebSocketHandlerAdapter。最后一部分是让 WebSocketHandlerAdapter 理解传入的反应式运行时请求。由于使用的是 Netty(默认配置)，我们需要使用 ReactorNettyRequestUpgradeStrategy 对象配置 WebSocketService。

```java
package com.apress.springbootrecipes.echo;

import org.springframework.boot.SpringApplication;
import org.springframework.boot.autoconfigure
  .SpringBootApplication;
import org.springframework.context.annotation.Bean;
import org.springframework.core.Ordered;
import org.springframework.web.reactive.HandlerMapping;
import org.springframework.web.reactive.handler
  .SimpleUrlHandlerMapping;
import org.springframework.web.reactive.socket.WebSocketHandler;
import org.springframework.web.reactive.socket.server
  .WebSocketService;
import org.springframework.web.reactive.socket.server.support.
HandshakeWebSocketService;
import org.springframework.web.reactive.socket.server.support.
WebSocketHandlerAdapter;
import org.springframework.web.reactive.socket.server.upgrade.
ReactorNettyRequestUpgradeStrategy;

import java.util.HashMap;
import java.util.Map;
```

```java
@SpringBootApplication
public class EchoApplication {

  public static void main(String[] args) {
      SpringApplication.run(EchoApplication.class, args);
  }

  @Bean
  public EchoHandler echoHandler() {
      return new EchoHandler();
  }

  @Bean
  public HandlerMapping handlerMapping() {
    Map<String, WebSocketHandler> map = new HashMap<>();
    map.put("/echo", echoHandler());

    var mapping = new SimpleUrlHandlerMapping();
    mapping.setUrlMap(map);
    mapping.setOrder(Ordered.HIGHEST_PRECEDENCE);
    return mapping;
  }

  @Bean
  public WebSocketHandlerAdapter handlerAdapter() {
      return new WebSocketHandlerAdapter(webSocketService());
  }

  @Bean
  public WebSocketService webSocketService() {
      var strategy = new ReactorNettyRequestUpgradeStrategy();
      return new HandshakeWebSocketService(strategy);
  }
}
```

现在可以启动服务器并使用 WebFlux 处理 WebSocket。

1. 用 HTML 和 JavaScript 编写客户端

现在服务器已经就绪，我们需要一个客户端来连接到 WebSocket 端点。为此，需要编写一些 JavaScript 和 HTML 代码。编写以下 app.js 文件并将其放到 src/main/resources/static 目录中。

```javascript
var ws = null;
var url = "ws://localhost:8080/echo";

function setConnected(connected) {
    document.getElementById('connect').disabled = connected;
    document.getElementById('disconnect').disabled = !connected;
    document.getElementById('echo').disabled = !connected;
}

function connect() {
    ws = new WebSocket(url);

    ws.onopen = function () {
        setConnected(true);
    };

    ws.onmessage = function (event) {
        log(event.data);
    };

    ws.onclose = function (event) {
        setConnected(false);
        log('Info: Closing Connection.');
    };
}
function disconnect() {
    if (ws != null) {
        ws.close();
        ws = null;
    }
    setConnected(false);
}
function echo() {
```

```
    if (ws != null) {
      var message = document.getElementById('message').value;
      log('Sent: ' + message);
      ws.send(message);
    } else {
        alert('connection not established, please connect.');
    }
}

function log(message) {
  var console = document.getElementById('logging');
  var p = document.createElement('p');
  p.appendChild(document.createTextNode(message));
  console.appendChild(p);
  while (console.childNodes.length > 12) {
      console.removeChild(console.firstChild);
  }
  console.scrollTop = console.scrollHeight;
}
```

上述代码中有多个函数。第一个函数是 connect，按下 index.html 页面上的 Connect 按钮时将触发该函数，这将打开到 ws://localhost:8080/echo 的 WebSocket 连接，服务器侧触发的处理器是前面创建和注册的 EchoHandler。连接到服务器将创建一个 WebSocket JavaScript 对象，这将使你能够侦听客户端上的消息和事件。这里定义了 onopen、onmessage 和 onclose 回调函数。最重要的是 onmessage，因为只要消息从服务器传入，就会调用 onmessage 函数。该函数只调用 log 函数，log 函数将接收到的消息添加到 logging 元素中并输出到屏幕上。

接下来是 disconnect 函数，该函数负责关闭 WebSocket 连接并清理 JavaScript 对象。最后，还有一个 echo 函数，每当按下 Echo message 按钮时，就会调用这个函数向服务器发送一条指定的消息(并最终被服务器发送回来)。

要使用 app.js 文件中的函数，添加如下代码所示的 index.html 文件。

```
<!DOCTYPE html>
<html>
<head>
    <link type="text/css" rel="stylesheet" href="https://cdnjs.cloudflare.com/ajax/libs/semantic-ui/2.2.10/semantic.min.css" />
    <script type="text/javascript" src="app.js"></script>
</head>
```

```html
<body>
<div>
  <div id="connect-container" class="ui centered grid">
    <div class="row">
        <button id="connect" onclick="connect();" class="ui green button ">Connect</button>
        <button id="disconnect" disabled="disabled" onclick="disconnect();" class="ui red button">Disconnect</button>
    </div>
    <div class="row">
        <textarea id="message" style="width: 350px" class="ui input" placeholder="Message to Echo"></textarea>
    </div>
    <div class="row">
        <button id="echo" onclick="echo();" disabled="disabled" class="ui button">Echo message</button>
    </div>
  </div>
  <div id="console-container">
    <h3>Logging</h3>
    <div id="logging"></div>
  </div>
</div>
</body>
</html>
```

现在，部署好应用程序后，客户端可以连接到已编写的 WebSocket 服务，向服务器发送一些消息并接收服务器的反馈，如图 5-5 所示。

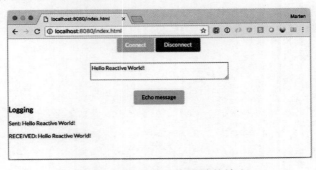

图 5-5　WebSocket 客户端的输出

2. 编写集成测试

编写集成测试需要做一些工作，包括与服务器建立 WebSocket 连接，手动发送消息并检查响应。但在此之前，需要先启动服务器。

要启动服务器，需要使用@RunWith（SpringRunner.class)和@SpringBootTest (webEnvironment = SpringBootTest.WebEnvironment.RANDOM_PORT)注解一个类。我们实际上想启动应用程序，而不是默认的 WebEnvironment.MOCK 环境。在 int 字段上添加@LocalServerPort 注解来获取实际的端口号。构造要连接的目标 URI 需要这个端口号。

```
package com.apress.springbootrecipes.echo;

@RunWith(SpringRunner.class)
@SpringBootTest(webEnvironment = SpringBootTest.WebEnvironment
  .RANDOM_PORT)
public class EchoHandlerIntegrationTest {
  @LocalServerPort
  private int port;
}
```

使用默认的 Java WebSocket API 测试 WebSocket 相当容易。为了测试，可以编写一个基本的 WebSocket 客户端来记录接收到的消息和会话。为此，需要一个 WebSocket 客户端，例如 Tomcat。

```
<dependency>
  <groupId>org.apache.tomcat.embed</groupId>
  <artifactId>tomcat-embed-websocket</artifactId>
  <scope>test</scope>
</dependency>
```

然而，添加此客户端将导致启动的服务器是 Tomcat，而不是我们希望的反应式的 Netty。在 EchoHandlerIntegrationTest 类中添加以下代码可以规避这个问题。

```
public class EchoHandlerIntegrationTest {

// Remainder of class omitted

  @Configuration
  @Import(EchoApplication.class)
  public static class EchoHandlerIntegrationTestConfiguratation {
```

```
    @Bean
    public NettyReactiveWebServerFactory webServerFactory() {
      return new NettyReactiveWebServerFactory();
    }
  }
}
```

这个新增的@Configuration 类将显式地启动 Netty 而不是 Tomcat。通过这种方式，我们就可以获得像正常运行应用程序那样的测试环境。取决于所使用的客户端(或测试方式)，可能没必要这样做。

现在编写 SimpleTestClientEndpoint 类，并使用@ClientEndpoint 进行注解；该类中的方法分别用@OnOpen、@OnClose 和@OnMessage 注解。在连接打开、关闭和收到消息时，应用程序将分别回调相应的方法。接下来有两个辅助方法：sendTextAndWait 和 closeAndWait。sendTextAndWait 方法使用 Session 发送消息并等待响应。closeAndWait 方法关闭会话并等待确认。最后，SimpleTestClientEndpoint 类提供一些 getter 方法来接收状态并在测试方法中进行验证。

```
@ClientEndpoint
public static class SimpleTestClientEndpoint {

  private List<String> received = new ArrayList<>();
  private Session session;
  private CloseReason closeReason;
  private boolean closed = false;

  @OnOpen
  public void onOpen(Session session) {
      this.session = session;
  }

  @OnClose
  public void onClose(Session session, CloseReason reason) {
    this.closeReason = reason;
    this.closed = true;
  }

  @OnMessage
  public void onMessage(String message) {
      this.received.add(message);
```

```java
  }

  public void sendTextAndWait(String text, long timeout)
      throws IOException, InterruptedException {
    var current = received.size();
    session.getBasicRemote().sendText(text);
    wait(() -> received.size() == current, timeout);
  }

  public void closeAndWait(long timeout)
        throws IOException, InterruptedException {
    if (session != null && !closed) {
      session.close();
    }
    wait(() -> closeReason == null, timeout);
  }

  private void wait(Supplier<Boolean> condition, long timeout)
          throws InterruptedException {
    var waited = 0;
    while (condition.get() && waited < timeout) {
      Thread.sleep(1);
      waited += 1;
    }
  }

  public CloseReason getCloseReason() {
    return closeReason;
  }

  public List<String> getReceived() {
      return this.received;
  }

  public boolean isClosed() {
      return closed;
  }
}
```

在编写好上述辅助方法之后,接下来开始编写测试代码。使用 WebSocket 库中的 ContainerProvider 类来获取容器。在建立实际连接之前,必须使用 port 字段的值构造 URI。接下来,将 SimpleTestClientEndpoint 类的实例 testClient 作为参数,通过容器的 connectToServer 方法连接到服务器。建立连接后,testClient 向服务器发送一条文本消息并等待一段时间,然后关闭连接(仅仅是为了清理资源)。最后,对接收到的消息进行验证。

```java
@Test
public void sendAndReceiveMessage() throws Exception {
  var container = ContainerProvider.getWebSocketContainer();
  var uri = URI.create("ws://localhost:" + port + "/echo");
  var testClient = new SimpleTestClientEndpoint();
  container.connectToServer(testClient, uri);

  testClient.sendTextAndWait("Hello World!", 200);
  testClient.closeAndWait(2);

  assertThat(testClient.getReceived())
          .containsExactly("RECEIVED: Hello World!");
}
```

第 6 章

Spring Security 介绍

本章介绍 Spring Boot 中的 Spring Security[1] 集成。Spring Security 用于应用程序的认证和授权。Spring Security 采用插件化的机制支持认证和授权处理过程，默认情况下支持不同的机制。对于认证机制，Spring Security 以开箱即用的方式支持 JDBC、LDAP 和属性文件。

6.1 在 Spring Boot 应用程序中启用安全特性

6.1.1 问题

你有一个基于 Spring Boot 的应用程序，并希望在这个应用程序中启用安全特性。

6.1.2 解决方案

添加 spring-boot-starter-security 依赖项，以便 Spring Boot 为应用程序自动配置和设置安全功能。

6.1.3 工作原理

首先，需要将 Spring Security 库放置到应用程序中，要做到这一点，需要将 spring-boot-starter-security 添加到依赖关系清单中。

```
<dependency>
  <groupId>org.springframework.boot</groupId>
  <artifactId>spring-boot-starter-security</artifactId>
</dependency>
```

---

[1] https://projects.spring.io/spring-security/.

这将为项目添加 spring-security-core、spring-security-config 和 spring-security-web 依赖项。Spring Boot 检测到这些 JAR 文件中包含了某些类，从而自动启用安全特性。

Spring Boot 将按照以下方式配置 Spring Security：具有基本身份验证和表单登录的身份验证功能；启用用于安全特性的 HTTP 头；Servlet API 集成；匿名登录功能以及禁用资源缓存功能。

■ **警告：**
Spring Boot将添加一个默认用户，用户名是user，带有随机生成的密码，启动日志记录了相关信息。这个用户名和密码仅用于测试、原型开发或演示，不要在真实的系统中使用这个系统生成的用户！

在 3.2 小节的依赖项清单中添加 spring-boot-starter-security 依赖项时，Spring Boot 将自动保护所有对外公开的端点。在启动应用程序时，系统生成的密码将记录在日志中，如图 6-1 所示。

```
2018-09-23 19:58:38.882  INFO 58204 --- [           main] .s.s.UserDetailsServiceAutoConfiguration :

Using generated security password: 033397ce-d724-4c7a-a717-edab7747f99d

2018-09-23 19:58:38.932 DEBUG 58204 --- [           main] s.s.c.a.w.c.WebSecurityConfigurerAdapter :
```

图 6-1　日志中记录的系统生成的密码

Spring Boot 提供一些属性来配置默认用户，这些属性位于 spring.security 名称空间中，如表 6-1 所示。

表 6-1　配置默认用户的属性

属性	说明
spring.security.user.name	默认用户名称，默认为 user
spring.security.user.password	默认用户密码，默认为系统产生的 UUID
spring.security.user.roles	默认用户角色。默认未设置

在添加依赖项并启动 LibraryApplication 之后，端点已受到了安全保护。当试图从 http://localhost:8080/books 获取图书列表时，结果将是一个状态为 401-Unauthorized 的 HTTP 响应，如图 6-2 所示。

在添加正确的身份验证头(用户名 user、日志中的密码或 spring.security.user.password 属性中指定的密码)时，返回的结果将是常规的图书列表，如图 6-3 所示。

```
code/ch06/recipe_6_1_i  master x
▶ http :8080/books
HTTP/1.1 401
Cache-Control: no-cache, no-store, max-age=0, must-revalidate
Content-Type: application/json;charset=UTF-8
Date: Sun, 23 Sep 2018 18:08:46 GMT
Expires: 0
Pragma: no-cache
Set-Cookie: JSESSIONID=EB5370933EB2198BE3B766B4F9A2C339; Path=/; HttpOnly
Transfer-Encoding: chunked
WWW-Authenticate: Basic realm="Realm"
X-Content-Type-Options: nosniff
X-Frame-Options: DENY
X-XSS-Protection: 1; mode=block

{
    "error": "Unauthorized",
    "message": "Unauthorized",
    "path": "/books",
    "status": 401,
    "timestamp": "2018-09-23T18:08:46.032+0000"
}
```

图 6-2　未经认证访问的结果

```
~/Repositories/spring-boot-recipes  master x
▶ http -a user:033397ce-d724-4c7a-a717-edab7747f99d :8080/books
HTTP/1.1 200
Cache-Control: no-cache, no-store, max-age=0, must-revalidate
Content-Type: application/json;charset=UTF-8
Date: Mon, 24 Sep 2018 17:41:50 GMT
Expires: 0
Pragma: no-cache
Set-Cookie: JSESSIONID=D434219EEF940FB083C1DC2B44BB1717; Path=/; HttpOnly
Transfer-Encoding: chunked
X-Content-Type-Options: nosniff
X-Frame-Options: DENY
X-XSS-Protection: 1; mode=block

[
    {
        "authors": [
            "J.R.R. Tolkien"
        ],
        "isbn": "9780618260300",
        "title": "The Hobbit"
    },
    {
        "authors": [
            "George Orwell"
        ],
        "isbn": "9780451524935",
        "title": "1984"
    },
```

图 6-3　通过认证的访问结果

1. 测试安全性

当通过@WebMvcTest 注解使用 Spring Security 对端点进行保护时，Spring Boot 将自动应用安全框架。Spring Security 提供了一些非常便于使用的注解(参见表 6-2)，使用这些注解有助于编写测试。

■ 注意：
如果希望测试没有安全保护的控制器程序，可以通过将@WebMvcTest的secure属性设置为false来禁用安全保护，该属性的默认值为true。

表 6-2　Spring Security 中用于测试的注解

注解	说明
@WithMockUser	使用指定了用户名、密码、角色/授权的用户运行测试
@WithAnonymousUser	使用匿名用户运行测试
@WithUserDetails	使用配置了用户名的用户运行测试，通过 UserDetailsService 接口查询用户名

要使用这些注解，必须添加 spring-security-test 依赖项。

```xml
<dependency>
  <groupId>org.springframework.security</groupId>
  <artifactId>spring-security-test</artifactId>
  <scope>test</scope>
</dependency>
```

有了这个依赖项，可以对 3.2 小节中的 BookControllerTest 类进行扩展和修改。如果你不介意为了测试禁用安全性，可以向测试类添加@WebMvcTest(value = BookController.class, secure = false)注解。添加该注解之后，Spring Boot 将不会添加安全过滤器，因此安全特性将被禁用。测试将会成功运行。

```
@RunWith(SpringRunner.class)
@WebMvcTest(value = BookController.class, secure = false)
public class BookControllerUnsecuredTest { ... }
```

但是，如果希望在启用了安全特性的情况下进行测试，则需要对测试类进行一些微小的修改。首先，添加@WithMockUser 注解以便使用经过认证的用户来测试应用程序。其次，由于 Spring Security 默认情况下启用了 CSRF[2] 保护，因此需要在请求消息中添加一个消息头或参数。当使用 Mock MVC 框架时，Spring Security 为此提供了一个 RequestPostProcessor 类型的实例，即 CsrfRequestPostProcessor。SecurityMockMvcRequestPostProcessors 类包含的工厂方法可以很容易地使用 CsrfRequestPostProcessor 实例。

```
@RunWith(SpringRunner.class)
@WebMvcTest(BookController.class)
```

---

[2] Cross Site Request Forgery，即跨站点请求伪造。

```
@WithMockUser
public class BookControllerSecuredTest {

  @Test
  public void shouldAddBook() throws Exception {

    when(bookService.create(any(Book.class))).thenReturn(new
    Book("123456789", "Test Book Stored", "T. Author"));

    mockMvc.perform(post("/books")
        .contentType(MediaType.APPLICATION_JSON)
        .content("{ \"isbn\" : \"123456789\"}, \"title\" : \"Test
        Book\", \"authors\" : [\"T. Author\"]")
        .with(csrf()))
        .andExpect(status().isCreated())
        .andExpect(header()
          .string("Location","http://localhost/books/123456789"));
  }
}
```

现在这个测试使用@WithMockUser 中指定的用户，在这里它使用默认的用户配置，用 user 作为用户名，password 作为密码。包含 with(csrf())的代码行负责将 CSRF 令牌添加到请求中。

测试时使用哪个选项取决于测试的需求。例如，需要在控制器中针对当前用户进行测试，则通常应该启用安全保护功能，并且应该使用@WithMockUser 或@WithUserDetails 指定用户。如果不需要针对某个用户进行测试，并且可以在没有安全保护功能的情况下测试控制器(同时也没有附加的安全规则，参见 6.2 小节)，那么在运行应用程序时可以禁用安全特性。

2. 安全特性的集成测试

是否能使用@SpringBootTest 编写集成测试取决于是否使用@With*注解。对于默认的模拟环境，仍然可以使用@With*注解。

```
@RunWith(SpringRunner.class)
@SpringBootTest
@WithMockUser
@AutoConfigureMockMvc
public class BookControllerIntegrationMockTest { ... }
```

这个测试将创建一个几乎完全成熟的应用程序，但是仍然使用 Mock MVC 来访问端点。它的运行过程与前面的测试完全相同,这也是@WithMockUser 和 with(csrf()) 仍然起作用的原因。如果测试访问的是外部端口,上述代码将无法正常运行。

要在一个端口上测试应用程序,需要通过 TestRestTemplate 和/或 WebTestClient 类型的测试客户端运行测试,并在请求消息中传递用于认证的消息头,或者首先在集成测试中执行基于表单的登录来实现测试流程。要编写成功的集成测试,需要注入 TestRestTemplate 类;并且在向服务器发出实际的请求之前,需要使用 withBasicAuth 辅助方法设置基本的身份验证消息头。

■ 提示:
在编写上述测试和使用默认用户时,你可能希望使用spring .security.user.password 属性设置一个默认密码。本节使用@TestPropertySource注解来完成这项工作,但是你也可以在application.properties文件中添加该属性。

```
@RunWith(SpringRunner.class)
@SpringBootTest(webEnvironment = SpringBootTest.WebEnvironment
.RANDOM_PORT)
@TestPropertySource(properties = "spring.security.user
.password=s3cr3t")
public class BookControllerIntegrationTest {

  @Autowired
  private TestRestTemplate testRestTemplate;

  @MockBean
  private BookService bookService;

  @Test
  public void shouldReturnListOfBooks() throws Exception {
    when(bookService.findAll()).thenReturn(Arrays.asList(
      new Book("123","Spring 5 Recipes","Marten Deinum","Josh Long"),
      new Book("321", "Pro Spring MVC", "Marten Deinum","Colin
      Yates")));
    ResponseEntity<Book[]> books = testRestTemplate
      .withBasicAuth("user", "s3cr3t")
      .getForEntity("/books", Book[].class);

    assertThat(books.getStatusCode()).isEqualTo(HttpStatus.OK);
```

```
        assertThat(books.getBody()).hasSize(2);
    }
}
```

这个测试使用默认配置的 TestRestTemplate 的实例来发出请求。withBasicAuth 使用默认的用户 user 和预设的 s3cr3t 字符串作为用户名和密码发送到服务器。可以使用 getForEntity 获得服务器返回的结果,其中包括有关响应的一些附加信息。使用 ResponseEntity 还可以验证状态代码等。

当测试基于 WebFlux 的应用程序而不是 TestRestTemplate 时,你将需要用到 WebTestClient(关于 Spring WebFlux 的更多信息,请参见第 5 章)。WebTestClient 的 headers()函数可以向请求添加额外的消息头。它公开了 HttpHeaders 类,而 HttpHeaders 类提供了 setBasicAuth 方法,使用该方法设置基本的身份验证非常方便。

```java
@RunWith(SpringRunner.class)
@SpringBootTest(webEnvironment=SpringBootTest.WebEnvironment.
    RANDOM_PORT)
@TestPropertySource(properties="spring.security.user.password=
    s3cr3t")
public class BookControllerIntegrationWebClientTest {

    @Autowired
    private WebTestClient webTestClient;

    @MockBean
    private BookService bookService;

    @Test
    public void shouldReturnListOfBooks() throws Exception {

      when(bookService.findAll()).thenReturn(Arrays.asList(
        new Book("123","Spring 5 Recipes", "Marten Deinum", "Josh Long"),
        new Book("321", "Pro Spring MVC", "Marten Deinum", "Colin
            Yates")));
    webTestClient
          .get()
            .uri("/books")
            .headers(headers->headers.setBasicAuth("user","s3cr3t"))
          .exchange()
            .expectStatus().isOk()
```

```
            .expectBodyList(Book.class).hasSize(2);
    }
}
```

代码中使用 exchange()函数构建和触发请求,然后验证结果的状态码是 HTTP 200 (OK),消息中包含两本书的信息。

6.2 登录 Web 应用

6.2.1 问题

一个安全的应用程序要求用户在访问某些受安全保护的功能之前先登录。这对于在开放的互联网上运行的应用程序尤其重要,因为,如果不用登录就可以访问,那么黑客就可以很轻松地访问这些受保护的功能。大多数应用程序都必须提供某种方式,让用户输入其凭据以便登录。

6.2.2 解决方案

Spring Security 支持多种方式供用户登录到 Web 应用中。它支持基于表单的登录,并提供默认的登录表单网页。你还可以提供一个自定义网页作为登录页面。此外,Spring Security 还通过处理 HTTP 请求头部字段中提供的基本身份验证凭据来支持 HTTP 基本身份验证。HTTP 基本身份验证还可用于对远程处理协议和 Web 服务发出的请求进行身份验证。

应用程序的某些部分可能允许匿名访问(例如,访问欢迎页面)。Spring Security 提供了匿名登录服务,可以为匿名用户分配主体并授予权限,因此在定义安全策略时,你可以像处理普通用户一样处理匿名用户。

Spring Security 还支持"记住我"的登录方式,它可以跨多个浏览器会话记住用户的身份,这样用户第一次登录后就不必再次登录。

6.2.3 工作原理

当找不到显式的 WebSecurityConfigurerAdapter 实例时,Spring Boot 启用默认安全设置。如果能找到一个或多个 WebSecurityConfigurerAdapter 实例,它将使用这些实例来配置安全特性。

为了帮助你更好地单独理解各种登录机制,让我们首先禁用默认的安全配置。

> ■ 提醒:
> 我们通常希望尽可能使用默认值,只禁用不需要的内容,例如使用 httpBasic().disable()禁用基本的HTTP身份验证,而不是禁用所有安全默认设置!

```
@Configuration
public class LibrarySecurityConfig extends
  WebSecurityConfigurerAdapter {

    public LibrarySecurityConfig() {
        super(true); // disable default configuration
    }
}
```

注意，如果启用 HTTP 自动配置，那么接下来介绍的登录服务将会自动注册。但是，如果禁用默认配置或要自定义这些服务，则必须显式配置相应的特性。

在启用身份验证功能之前，必须启用基本的 Spring Security 需求，至少需要配置异常处理和安全上下文集成。

```
@Override
protected void configure(HttpSecurity http) {

    http.securityContext()
           .and()
          .exceptionHandling();
}
```

没有这些基础设置，Spring Security 在登录后就不会存储用户信息，也不会为与安全相关的异常进行适当的异常转换(异常会向上层代码传递，这可能会将应用程序的一些内部信息暴露给外部代码)。你还可能希望启用 Servlet API 集成，以便可以使用 HttpServletRequest 类的方法在视图中执行检查。

```
@Override
protected void configure(HttpSecurity http) {
    http.servletApi();
}
```

1. HTTP 基础认证

可以通过 httpBasic()方法配置 HTTP 基础认证。当需要 HTTP 基础认证时，浏览器通常会显示一个登录对话框或浏览器特定的登录页面，供用户登录。

```
@Configuration
public class LibrarySecurityConfig extends
  WebSecurityConfigurerAdapter {

  @Override
```

```
protected void configure(HttpSecurity http) throws Exception {
    http
        ...
        .httpBasic();
}
```

2. 基于表单的登录

基于表单的登录服务将展现一个包含登录表单的网页，供用户输入登录详细信息并负责提交登录表单。通过 formLogin 方法配置基于表单的登录服务，如下所示。

```
@Configuration
public class LibrarySecurityConfig extends WebSecurityConfigurerAdapter {
 @Override
  protected void configure(HttpSecurity http) throws Exception {
    http
        ...
        .formLogin();
  }
}
```

默认情况下，Spring Security 会自动创建一个登录页面，并将其映射到地址 /login。因此，可以向应用程序添加一个链接(例如，3.3 小节的 index.html)，登录时请使用这个链接，如下所示：

```
<a th:href="/login" href="#">Login</a>
```

如果不喜欢默认的登录页面，可以提供自定义的登录页面。例如，使用 Thymeleaf 模板时可以在 src/main/resources/templates 目录下创建以下 login.html 文件。默认情况下，CSRF 保护是打开的，因此需要在表单中添加一个 CSRF 令牌。隐藏字段的作用正在于保存 CSRF 令牌。如果禁用了 CSRF(不推荐这样做)，应该删除相应的隐藏字段。

```
<!DOCTYPE html>
<html xmlns:th="http://www.thymeleaf.org">
<head>
    <title>Login</title>
    <link type="text/css" rel="stylesheet"
          href="https://cdnjs.cloudflare.com/ajax/libs/
          semantic-ui/ 2.2.10semantic.min.css">
```

```html
    <style type="text/css">
        body {
            background-color: #DADADA;
        }
        body > .grid {
            height: 100%;
        }
        .column {
            max-width: 450px;
        }
    </style>
</head>

<body>
<div class="ui middle aligned center aligned grid">
    <div class="column">
        <h2 class="ui header">Log-in to your account</h2>
        <form method="POST" th:action="@{/login}" class="ui large form">
            <input type="hidden"
                th:name="${_csrf.parameterName}" th:value="${_csrf.token}"/>
            <div class="ui stacked segment">
                <div class="field">
                    <div class="ui left icon input">
                        <i class="user icon"></i>
                        <input type="text" name="username" placeholder="Email address">
                    </div>
                </div>
                <div class="field">
                    <div class="ui left icon input">
                        <i class="lock icon"></i>
                        <input type="password" name="password"
                            placeholder="Password">
                    </div>
                </div>
                <button class="ui fluid large submit green button">Login</button>
```

```
            </div>
        </form>
    </div>
</div>
</body>
</html>
```

为了让 Spring Security 在请求登录时显示自定义的登录页面，必须在 loginPage 配置方法中指定其 URL。

```
@Configuration
public class LibrarySecurityConfig extends WebSecurityConfigurerAdapter{

    @Override
    protected void configure(HttpSecurity http) throws Exception {
        http
            ...
            .formLogin().loginPage("/login");
    }
}
```

最后，添加一个视图解析器来将地址/login 映射到 login.html 页面。为此，可以让 LibrarySecurityConfig 类实现 WebMvcConfigurer 接口并重写 addViewControllers 方法。

```
@Configuration
public class LibrarySecurityConfig extends WebSecurityConfigurerAdapter
                                    implements WebMvcConfigurer {
    ...

    public void addViewControllers(ViewControllerRegistry registry) {
        registry.addViewController("/login").setViewName("login");
    }
}
```

当用户请求一个安全的 URL 时，如果 Spring Security 显示登录页面，那么登录成功后，用户将被重定向到目标 URL。但是，如果用户直接通过登录页面的 URL 请求该页面，默认情况下，在成功登录后，用户会被重定向到上下文路径的根路径（即 http://localhost:8080/）。如果还没有在 Web 部署描述符中定义欢迎页面，你可能希望在登录成功时将用户重定向到默认的目标 URL。

```
@Configuration
public class LibrarySecurityConfig extends WebSecurityConfigurerAdapter
                                   implements WebMvcConfigurer {
    @Override
    protected void configure(HttpSecurity http) throws Exception {
        http
            ...
            .formLogin().loginPage("/login").defaultSuccessUrl
            ("/books");
    }
}
```

如果使用由 Spring Security 创建的默认登录页面,那么当登录失败时,Spring Security 将再次呈现登录页面,并显示错误信息。但是,如果指定了自定义登录页面,则必须通过 failureUrl() 函数配置登录出错时重定向的 URL。例如,可以再次重定向到自定义登录页面,并显示错误的请求参数。

```
@Configuration
public class LibrarySecurityConfig extends WebSecurityConfigurerAdapter
                                   implements WebMvcConfigurer {
    @Override
    protected void configure(HttpSecurity http) throws Exception {
        http
            ...
            .formLogin()
                .loginPage("/login.html")
                .defaultSuccessUrl("/books")
                .failureUrl("/login.html?error=true");
    }
}
```

然后登录页面应该测试错误的请求参数是否存在。如果发生错误,则必须通过访问会话作用域属性 SPRING_SECURITY_LAST_EXCEPTION 来显示错误消息,该属性存储了由 Spring Security 为当前用户引发的最后一个异常。

```
<form>
    ...
    <div th:if="${param.error}">
        <div class="ui error message" style="display: block;">
            Authentication Failed<br/>
```

```html
          Reason :
          <span th:text="${session.SPRING_SECURITY_LAST_
          EXCEPTION.message}" />
      </div>
  </div>

</form>
```

3. 注销服务

注销服务提供了一个处理器来处理注销请求。可以通过 logout()配置方法对其进行配置。

```java
@Configuration
public class LibrarySecurityConfig extends WebSecurityConfigurerAdapter
                                   implements WebMvcConfigurer {
  @Override
  protected void configure(HttpSecurity http) throws Exception {
      http
          ...
          .and()
              .logout();
  }
}
```

默认情况下，它被映射到/logout 地址，并且只对 POST 请求做出响应。可以在页面中添加一个简单的 HTML 表单来向用户提供注销功能。

```html
<form th:action="/logout" method="post"><button>Logout</button><form>
```

> **■ 注意：**
> 当使用CSRF保护时，不要忘了在表单中添加CSRF令牌，否则注销将失败。

默认情况下，当注销成功时，用户将被重定向到上下文路径的根目录，但有时可能希望将用户引导到另一个 URL，可以使用 logoutSuccessUrl()配置方法来完成这项操作。

```java
@Configuration
public class LibrarySecurityConfig extends WebSecurityConfigurerAdapter
                                   implements WebMvcConfigurer {
```

```
@Override
protected void configure(HttpSecurity http) throws Exception {
    http
        ...
        .and()
            .logout().logoutSuccessUrl("/");
    }
}
```

注销后，你可能会注意到，当使用浏览器的后退按钮时，仍然可以看到前面的页面，即使注销是成功的。这是因为浏览器默认情况下会缓存页面。为了避免出现这种情况，使用 headers() 配置方法启用安全头指示浏览器不要缓存页面即可。

```
@Configuration
public class LibrarySecurityConfig extends WebSecurityConfigurerAdapter
                                   implements WebMvcConfigurer {

    @Override
    protected void configure(HttpSecurity http) throws Exception {
        http
            ...
            .and()
                .headers();
    }
}
```

使用 headers() 配置方法后，不仅可以让浏览器不再缓存页面，还可以禁止内容嗅探并启用 x-frame 保护。启用这项保护后，使用浏览器后退按钮时，你将被再次重定向到登录页面。

4. 匿名登录

可以在 configure() 方法中通过 anonymous() 方法配置匿名登录服务，配置时可以自定义匿名用户的用户名和权限，匿名用户的默认用户名是 anonymousUser，默认权限是 ROLE_ANONYMOUS。

```
@Configuration
public class LibrarySecurityConfig extends WebSecurityConfigurerAdapter
                                   implements WebMvcConfigurer {

    @Override
```

```
protected void configure(HttpSecurity http) throws Exception {
    http
        ...
        .and()
        .anonymous().principal("guest").authorities("ROLE_GUEST");
    }
}
```

5. "记住我"功能

可以在configure()方法中通过rememberMe()方法配置"记住我"功能。默认情况下，该功能将用户名、用户密码、"记住我"的过期时间和私钥编码为一个令牌，并将该令牌存储在浏览器的cookie中。下次用户访问相同的Web应用时，浏览器将检测到这个令牌，以便用户能够自动登录。

```
@Configuration
public class LibrarySecurityConfig extends WebSecurityConfigurerAdapter
                                implements WebMvcConfigurer {

    @Override
    protected void configure(HttpSecurity http) throws Exception {
        http
            ...
            .and()
            .rememberMe();
    }
}
```

然而，这种静态的"记住我"令牌可能会导致安全问题，因为它们可能会被黑客捕获。Spring Security 支持滚动令牌以满足更高级的安全需求，但是这需要通过数据库使令牌持久化。有关滚动"记住我"令牌部署的详细信息，请参见 Spring Security 的参考文档。

6.3 用户认证

6.3.1 问题

当用户试图登录到应用程序以访问其安全资源时，必须对用户的主体进行身份验证并向该用户授予权限。

6.3.2 解决方案

在 Spring Security 中,身份验证由一个或多个 AuthenticationProvider 执行,多个 AuthenticationProvider 形成一个验证链。如果这些提供商程序中的任何一个成功地对用户进行了身份验证,则该用户将能够登录到应用程序中。如果任何提供商程序报告用户已被禁用或锁定,或凭据不正确,或者如果没有提供商程序能够验证用户,则用户将无法登录到此应用程序。

Spring Security 支持多种验证用户的方式,并为这些验证方式提供了内置的提供商程序实现,可以使用内置的 XML 元素轻松地配置这些提供商程序。大多数常见的身份验证提供商程序针对存储了用户详细信息的用户存储库(例如,应用程序的内存、关系数据库或 LDAP 存储库)对用户进行身份验证。

在存储库中存储用户详细信息时,应避免将用户密码以明文存储,因为这会使用户容易受到黑客攻击。相反,应该始终将加密的密码存储在存储库中。一种典型的方法是使用单向散列函数对密码进行编码。当用户输入登录密码时,对该密码应用相同的散列函数,并将结果与存储在存储库中的结果进行比较。Spring Security 支持多种密码编码算法(包括 BCrypt 和 SCrypt),并为这些算法提供内置的密码编码器。

6.3.3 工作原理

1. 使用内存内定义验证用户

如果应用程序的用户数量很少,而且很少修改他们的详细信息,那么可以考虑在 Spring Security 的配置文件中定义用户详细信息,以便将它们加载到应用程序的内存中。

```
@Configuration
public class LibrarySecurityConfig extends WebSecurityConfigurerAdapter{

...

  @Override
  protected void configure(AuthenticationManagerBuilder auth)
  throws Exception {
    UserDetails adminUser = User.withDefaultPasswordEncoder()
        .username("admin@books.io")
        .password("secret")
        .authorities("ADMIN","USER").build();

    UserDetails normalUser = User.withDefaultPasswordEncoder()
        .username("marten@books.io")
```

```
        .password("user")
        .authorities("USER").build();

UserDetails disabledUser = User.withDefaultPasswordEncoder()
        .username("marten@books.io")
        .password("user")
        .disabled(true)
        .authorities("USER").build();

auth.inMemoryAuthentication()
        .withUser(adminUser)
        .withUser(normalUser)
        .withUser(disabledUser);
    }
}
```

上述代码中，调用 User.withDefaultPasswordEncoder() 获得一个 UserDetails 实例，通过该实例可以使用加密密码构造用户。它将使用 Spring Security 的默认密码编码（默认情况下，使用 BCrypt 编码）。可以使用 inMemoryAuthentication() 方法添加用户详细信息；使用 withUser() 方法，可以定义用户。对于每个用户，可以指定用户名、密码、禁用状态和一组已授予的权限。禁用的用户无法登录到应用程序。

2. 根据数据库对用户进行身份验证

更典型的情况是，为了便于维护，应该将用户详细信息存储在数据库中。Spring Security 内置支持从数据库查询用户的详细信息。默认情况下，它使用以下 SQL 语句查询包含用户权限的用户详细信息：

```
SELECT username, password, enabled
FROM users
WHERE username = ?

SELECT username, authority
FROM authorities
WHERE username = ?
```

为了让 Spring Security 使用这些 SQL 语句查询用户详细信息，必须在数据库中创建相应的表。例如，可以使用以下 SQL 语句在数据库中创建数据表：

```
CREATE TABLE USERS (
    USERNAME VARCHAR(50) NOT NULL,
```

```
    PASSWORD VARCHAR(50)     NOT NULL,
    ENABLED  SMALLINT        NOT NULL,
    PRIMARY KEY (USERNAME)
);

CREATE TABLE AUTHORITIES (
    USERNAME     VARCHAR(50)   NOT NULL,
    AUTHORITY    VARCHAR(50)   NOT NULL,
    FOREIGN KEY (USERNAME) REFERENCES USERS
);
```

接下来,可以在这些表中存入一些用户详细信息,以便进行测试。这两个表格的数据见表 6-3 和表 6-4。

表 6-3 USERS 表的测试用户数据

用户名	密码	启用状态
admin@books.io	{noop}secret	1
marten@books.io	{noop}user	1
jdoe@books.net	{noop}unknown	0

■ 注意:

密码字段中的 {noop} 表示没有对存储的密码进行加密。Spring Security 使用委托来确定要使用哪种编码方法,可选的值包括{bcrypt}、{scrypt}、{pbkdf2}和{sha256}。保留{sha256}主要是出于兼容性原因,但该算法已不再安全。

表 6-4 AUTHORITIES 表的测试用户数据

用户名	权限
admin@books.io	ADMIN
admin@books.io	USER
marten@books.io	USER
jdoe@books.net	USER

为了让 Spring Security 访问这些表,必须声明一个数据源,以便能够连接到数据库。

在实现 LibrarySecurityConfig 类的 configure()方法时,使用 AuthenticationManager-Builder 实例的 jdbcAuthentication()配置方法,并将一个 DataSource 实例作为该方法的参数。通常,这个数据源是 Spring Boot 配置的 DataSource 实例,可以通过 spring.datasource 属性对其进行配置(更多详细信息,请参见 7.1 小节)。

```
@Configuration
public class LibrarySecurityConfig extends WebSecurityConfigurerAdapter{

    @Autowired
    private DataSource dataSource;

    @Override
    protected void configure(AuthenticationManagerBuilder auth)
    throws Exception {
        auth.jdbcAuthentication().dataSource(dataSource);
    }
}
```

但是，在某些情况下，应用程序可能已经在现有的数据库中定义了自己的用户存储库。例如，假设如下数据表是用以下 SQL 语句创建的，并且 MEMBER 表中的所有用户都是启用状态：

```
CREATE TABLE MEMBER (
    ID          BIGINT          NOT NULL,
    USERNAME    VARCHAR(50)     NOT NULL,
    PASSWORD    VARCHAR(32)     NOT NULL,
    PRIMARY KEY (ID)
);

CREATE TABLE MEMBER_ROLE (
    MEMBER_ID       BIGINT          NOT NULL,
    ROLE            VARCHAR(10)     NOT NULL,
    FOREIGN KEY (MEMBER_ID) REFERENCES MEMBER
);
```

假设将已有的用户数据存储在这些表中，如表 6-5 和表 6-6 所示。

表 6-5　MEMBER 表已有的用户数据

ID	用户名	密码
1	admin@ya2do.io	{noop}secret
2	marten@ya2do.io	{noop}user

表 6-6　MEMBER_ROLE 表已有的用户数据

MEMBER_ID	ROLE
1	ROLE_ADMIN
1	ROLE_USER
2	ROLE_USER

幸运的是，Spring Security 还支持使用自定义 SQL 语句来查询遗留数据库以获取用户详细信息。可以使用 usersByUsernameQuery() 和 authoritiesByUsernameQuery() 配置方法来指定查询用户信息和权限的语句。

```
@Configuration
public class LibrarySecurityConfig extends WebSecurityConfigurerAdapter{

    ...

    @Override
    protected void configure(AuthenticationManagerBuilder auth)
    throws Exception {
      auth.jdbcAuthentication()
        .dataSource(dataSource)
        .usersByUsernameQuery(
          "SELECT username, password, 'true' as enabled " +
          "FROM member WHERE username = ?")
        .authoritiesByUsernameQuery(
          "SELECT member.username, member_role.role as authorities " +
          "FROM member, member_role " +
          "WHERE member.username=?AND member.id=member_role.member_id");
    }
}
```

3. 加密密码

到目前为止，在存储用户详细信息时一直使用的是明文密码。但是这种方法很容易受到黑客攻击，所以应该先加密密码再存储它们。Spring Security 支持几种加密密码的算法。例如，可以选择 BCrypt 加密密码，这是一种单向散列算法。

■ 注意：
你可能需要一个辅助方法来计算密码的 BCypt 散列值，可以使用在线工具完成这项工作，例如 www.browserling.com/tools/bcrypt。或者可以简单地创建一个类，在类的 main 方法中调用 Spring Security 的 BCryptPasswordEncoder 方法。

当然，你必须将加密的密码，而不是明文密码，存储在数据库表中，如表 6-7 所示。要在 password 字段中存储 BCrypt 散列，该字段的长度不能小于 68 个字符(即 BCrypt 散列的长度+加密类型 {bcrypt}的长度)。

表 6-7 密码已加密的 USERS 表格中的测试用户数据

用户名	密码	启用状态
admin@ya2do.io	{bcrypt}$2a$10$E3mPTZb50e7sSW15fDx8Ne7hDZpfDjrmMPTTUp8wVjLTu.G5oPYCO	1
marten@ya2do.io	{bcrypt}$2a$10$5VWqjwoMYnFRTTmbWCRZT.iY3WW8ny27kQuUL9yPK1/WJcPcBLFWO	1
jdoe@does.net	{bcrypt}$2a$10$cFKh0.XCUOA9L.in5smIiO2QIOT8.6ufQSwIIC.AVz26WctxhSWC6	0

6.4 制定访问控制决策

6.4.1 问题

在身份验证过程中，应用程序将向已通过身份验证的用户授予一组权限。当此用户试图访问应用程序中的资源时，应用程序必须确定是否可以通过授予的权限或其他特性访问该资源。

6.4.2 解决方案

关于是否允许用户访问应用程序中的资源的决定称为访问控制决策。它是基于用户的认证状态、资源的性质以及访问属性而做出的决定。

6.4.3 工作原理

在启用 Spring Security 特性之后，可以使用 Spring 表达式语言(Spring Expression Language，SpEL)来创建强大的访问控制规则。Spring Security 支持一些开箱即用的表达式(见表 6-8)。使用诸如 and、or 和 not 之类的逻辑，可以创建非常强大和灵活的表达式。

表 6-8 Spring Security 内置的表达式

表达式	描述
hasRole('role') or hasAuthority('authority')	如果当前用户拥有指定的角色，则返回 true
hasAnyRole('role1', 'role2') / hasAnyAuthority('auth1', 'auth2')	如果当前用户拥有至少一个指定的角色，则返回 true

(续表)

表达式	描述
hasIpAddress('ip-address')	如果当前用户具有给定的 IP 地址，则返回 true
principal	表示当前用户
Authentication	对 Spring Security 身份验证对象的访问
permitAll	总是返回 true
denyAll	总是返回 false
isAnonymous()	如果当前用户是匿名用户，则返回 true
isRememberMe()	如果当前用户是通过"记住我"的功能登录的，则返回 true
isAuthenticated()	如果当前用户不是匿名用户，则返回 true
isFullyAuthenticated()	如果当前用户不是匿名用户，也不是通过"记住我"功能登录的，则返回 true

■ **警告：**

虽然角色和权限几乎是一样的，但在处理它们的方式上存在细微但重要的差异。当使用 hasRole 方法时，如果角色的传入值以 ROLE_(默认角色前缀)开头，则将检查该角色的权限；如果传入的值没有该前缀，则将在检查角色的权限之前添加该前缀。因此，调用 hasRole('ADMIN')将实际检查当前用户是否具有 ROLE_ADMIN 权限。当使用 hasAuthority 方法时，它将按参数的现有值进行检查。

如果某位用户具有 ADMIN 角色或是在本地计算机上登录的，以下表达式将授予该用户删除图书的权限。定义匹配器时，可以通过 access 方法而不是某个 has*方法来编写这样的表达式。

```
@Configuration
public class LibrarySecurityConfig extends WebSecurityConfigurerAdapter{

  ...

  @Override
  protected void configure(HttpSecurity http) throws Exception {
    http
      .authorizeRequests()
        .antMatchers(HttpMethod.GET,"/books*").hasAnyRole("USER",
          "GUEST")
        .antMatchers(HttpMethod.POST, "/books*").hasRole("USER")
        .antMatchers(HttpMethod.DELETE, "/books*")
```

```
            .access("hasRole('ROLE_ADMIN') or hasIpAddress('127.0.0.1')"+
                    "or hasIpAddress('0:0:0:0:0:0:0:1')")
            ...
    }
}
```

1. 使用表达式制定使用 Spring Bean 的访问控制策略

在表达式中使用@语法,可以调用应用程序上下文中的任意 bean。因此,你可以编写类似@accessChecker.hasLocalAccess(authentication)的表达式,并提供一个名为 accessChecker 的 bean,它具有 hasLocalAccess 方法,该方法接受 Authentication 对象作为参数。

```
package com.apress.springbootrecipes.library.security;

import org.springframework.security.core.Authentication;
import org.springframework.security.web.authentication.WebAuthenticationDetails;

@Component
public class AccessChecker {

    public boolean hasLocalAccess(Authentication authentication) {
        boolean access = false;
        if (authentication.getDetails() instanceof WebAuthenticationDetails) {
            WebAuthenticationDetails details =
                (WebAuthenticationDetails) authentication.getDetails();
            String address = details.getRemoteAddress();
            access = address.equals("127.0.0.1") ||
                     address.equals("0:0:0:0:0:0:0:1");
        }
        return access;
    }
}
```

AccessChecker 类与前面编写的自定义表达式处理器程序执行相同的检查,但不扩展 Spring Security 的类。

```
@Override
protected void configure(HttpSecurity http) throws Exception {
```

```
http.authorizeRequests()
    .antMatchers(HttpMethod,POST, "/books*").hasAuthority("USER")
    .antMatchers(HttpMethod.DELETE, "/books*")
      .access("hasAuthority('ADMIN') " +
         "or @accessChecker.hasLocalAccess(authentication)");
   ...
}
```

2. 用注解和表达式保护方法

可以使用@PreAuthorize 和@PostAuthorize 注解来保护方法的调用,而不只是保护 URL。在调用方法之前 Spring Security 会检查@PreAuthorize 注解中的条件是否满足,在调用方法之后会检查@PostAuthorize 注解的条件是否满足,并可以对条件表达式的返回值进行安全检查。使用这些注解,可以编写基于安全的表达式,就像使用基于 URL 的安全性一样。要启用注解处理,可以将 @EnableGlobalMethod-Security 注解添加到安全配置中,并将 prePostEnabled 属性设置为 true。

```
@Configuration
@EnableGlobalMethodSecurity(prePostEnabled = true)
public class LibrarySecurityConfig extends
  WebSecurityConfigurerAdapter {
... }
```

现在,可以使用@PreAuthorize 注解来保护应用程序了,如以下代码所示:

```
package com.apress.springbootrecipes.library;

import org.springframework.security.access.prepost.PreAuthorize;
import org.springframework.stereotype.Service;

import java.util.Map;
import java.util.Optional;
import java.util.concurrent.ConcurrentHashMap;

@Service
class InMemoryBookService implements BookService {

    private final Map<String, Book> books = new ConcurrentHashMap<>();

    @Override
```

```
    @PreAuthorize("isAuthenticated()")
    public Iterable<Book> findAll() {
        return books.values();
    }

    @Override
    @PreAuthorize("hasAuthority('USER')")
    public Book create(Book book) {
        books.put(book.getIsbn(), book);
        return book;
    }

    @Override
    @PreAuthorize("hasAuthority('ADMIN')or@accessChecker.hasLocalAccess
    (authentication)")
    public void remove(Book book) {
      books.remove(book.getIsbn());
    }

    @Override
    @PreAuthorize("isAuthenticated()")
      public Optional<Book> find(String isbn) {
        return Optional.ofNullable(books.get(isbn));
    }
}
```

@PreAuthorize 注解将触发 Spring Security 验证其所包含的表达式。如果表达式通过验证,那么访问获得授权,否则代码将抛出一个异常并向用户表明他没有访问权限。

6.5 向 WebFlux 应用程序添加安全特性

6.5.1 问题

你有一个使用 Spring Web Flux 构建的应用程序(参见第 5 章),希望使用 Spring Security 保护它。

6.5.2 解决方案

当在基于 WebFlux 的应用程序中添加 Spring Security 的依赖项之后，Spring Boot 将自动启用安全性。Spring Boot 将向应用程序添加一个 @EnableWebFluxSecurity 注解的配置类。该注解将导入默认的 Spring Security WebFluxSecurityConfiguration 类。

6.5.3 工作原理

Spring WebFlux 应用程序与普通的 Spring MVC 应用程序在本质上是非常不同的；尽管如此，Spring Boot 和 Spring Security 力求使构建安全的基于 WebFlux 的应用程序变得更容易。

以 5.3 小节创建的 WebFlux 应用程序为例，为了启用安全性，需要将 spring-boot-starter-security 添加到该应用程序的依赖项清单中。

```
<dependency>
  <groupId>org.springframework.boot</groupId>
  <artifactId>spring-boot-starter-security</artifactId>
</dependency>
```

上述代码将为项目添加 spring-security-core、spring-security-config 和 spring-security-web 依赖项。Spring Boot 检测到这些 JAR 文件中某些类的存在，从而自动启用安全特性。

Spring Boot 为 Spring Security 启用以下配置：使用基本身份验证和表单登录进行身份验证，启用 HTTP 安全头字段，并要求访问任何资源之前必须先登录。

■警告：
Spring Boot 将添加一个默认用户，用户名为 user，并带有系统生成的密码，在启动日志中可以查看此用户的信息。该用户仅用于测试、原型制作或演示，不要在真实的系统中使用这个系统生成的用户！

当在 5.3 小节所述的依赖项清单中添加 spring-boot-starter-security 依赖项后，Spring Security 将自动保护所有暴露的端点。在启动应用程序时，可以在日志中看到系统生成的用户名和密码，如图 6-4 所示。

图 6-4　安全的 WebFlux 应用程序的启动日志

在尝试访问 http://localhost:8080/地址时，浏览器将显示一个登录页面，如图 6-5 所示。

图 6-5　默认登录页面

1. 保护 URL 访问

可以通过添加自定义 SecurityWebFilterChain 类来配置访问规则。首先，让我们创建 OrdersSecurityConfiguration 类。

```
@Configuration
public class OrdersSecurityConfiguration { ... }
```

来自 Spring Security 的 WebFluxSecurityConfiguration 类检测 SecurityWebFilterChain 的实例(其中包含了安全配置)，该实例被封装为一个 WebFilter 实例，而 WebFlux 则使用它对传入的请求做出访问权限决策(就像普通的 Servlet 过滤器那样)。

目前，配置只是启用了安全特性，让我们添加一些安全规则。

```
@Bean
SecurityWebFilterChain springWebFilterChain(ServerHttpSecurity http)
throws Exception {
    return http
            .authorizeExchange()
            .pathMatchers("/").permitAll()
            .pathMatchers("/orders*").hasRole("USER")
            .anyExchange().authenticated()
            .and().build();
}
```

ServerHttpSecurity 看起来与 6.4 小节中的 HttpSecurity 有点相似，用于添加安全规则并进行进一步的配置(例如添加/删除消息头和配置登录方法)。上述代码通过 authorizeExchange()方法编写如下规则：使用安全的地址，每个人都可以访问根目录/，但只有角色是 USER 的用户才能访问/orders 地址。对于其他请求，用户必须至少先经过身份验证。最后，需要调用 build()方法来实际构建 SecurityWebFilterChain 对象。

在 authorizeExchange()方法之后还可以使用 headers()配置方法向请求添加安全头，使用 csrf()添加 CSRF 保护等。

2. 登录 Web Flux 应用程序

可以通过显式配置来覆盖部分默认配置，可以覆盖要使用的身份验证管理器和用于存储安全上下文的存储库。Spring Boot 会自动检测身份验证管理器，只需要注册类型为 ReactiveAuthenticationManager 或 UserDetailsRepository 的 bean 即可。

还可以通过配置 ServerSecurityContextRepository 来配置存储 SecurityContext 的位置。使用的默认实现是 WebSessionServerSecurityContextRepository，它将上下文存储在 WebSession 中。另一个默认实现是 NoOpServerSecurityContextRepository，用于无状态应用程序。

```
@Bean
SecurityWebFilterChain springWebFilterChain(HttpSecurity http) throws
Exception {
    return http
            .httpBasic().
            .and().formLogin().
        .authenticationManager(new CustomReactiveAuthenticationManager())
        .securityContextRepository(
            new ServerWebExchangeAttributeSecurityContextRepository())
```

```
            .and().build();
}
```

上述代码使用 CustomReactiveAuthenticationManager 和无状态 NoOpServerSecurityContextRepository 覆盖默认值。但是，对于我们的应用程序，我们将沿用默认配置。

3. 身份验证

在基于 Spring WebFlux 的应用程序中通过 ReactiveAuthenticationManager 对用户进行身份验证，这是只有 authenticate 方法的接口。你既可以提供自己的实现，也可以使用提供的两个实现之一。第一个实现是 UserDetailsRepositoryAuthenticationManager 类，它封装了一个 ReactiveUserDetailsService 实例。

■ 注意：
ReactiveUserDetailsService只有一个实现，即MapReactiveUserDetailsService，它是内存内实现。你可以基于反应式数据存储(如MongoDB或Couchbase)提供自己的实现。

另一个实现ReactiveAuthenticationManagerAdapter 实际上是常规AuthenticationManager 的封装器。它将封装一个常规的实例，因此可以以一种反应式的方式使用阻塞式实现。这并不能使它们变成反应式的：它们仍然会阻塞，但可以通过这种方式重用。有了这个封装器，还可以将 JDBC、LDAP 等用于反应式应用程序。

当在 Spring WebFlux 应用程序中配置 Spring Security 时，可以将 ReactiveAuthenticationManager 类的实例添加到 Java 配置类或 UserDetailsRepository 类中。当检测到后者时，它将自动封装在 UserDetailsRepositoryAuthenticationManager 类中。

```
@Bean
public MapUserDetailsRepository userDetailsRepository() {
    UserDetails marten =
      User.withUsername("marten").password("secret")
      .roles("USER").build();
    UserDetails admin =
      User.withUsername("admin").password("admin")
          .roles("USER","ADMIN").build();
    return new MapUserDetailsRepository(marten, admin);
}
```

现在运行应用程序，你可以自由访问根目录/页面，但当访问以/orders 开头的 URL 时，浏览器会显示一个登录表单，如图 6-5 所示。当输入某个预定义用户的凭据时，应用程序应该允许你访问请求的 URL。

4. 制定访问控制策略

应用程序在运行的某个时刻需要根据用户拥有的权限或角色授予用户访问权限。Spring Security 提供了一些内置表达式来处理这个问题，如表 6-9 所示。

表 6-9　Spring Security WebFlux 内置表达式

表达式	描述
hasRole('role') or hasAuthority('authority')	如果当前用户具有给定的角色，则返回 true
permitAll()	总是返回 true
denyAll()	总是返回 false
authenticated()	如果用户通过身份验证，则返回 true
access()	使用一个函数来确定是否授予访问权限

■ **警告：**

虽然角色和权限几乎相同，但在处理它们的方式上存在细微但重要的差异。当使用 hasRole 方法时，如果角色以传入值 ROLE_(默认角色前缀)开头，则将检查该角色的权限；如果传入的值没有该前缀，则将在检查角色的权限之前添加该前缀。因此，调用 hasRole('ADMIN') 将实际检查当前用户是否具有 ROLE_ADMIN 权限。当使用 hasAuthority 方法时，它将按参数的现有值进行检查。

```
@Bean
SecurityWebFilterChain springWebFilterChain(HttpSecurity http) throws
Exception {
    return http
        .authorizeExchange()
            .pathMatchers("/").permitAll()
            .pathMatchers("/orders*").access(this::ordersAllowed)
            .anyExchange().authenticated()
        .and()
        .build();
}
private Mono<AuthorizationDecision> ordersAllowed(Mono<Authentication>
authentication, AuthorizationContext context) {
    return authentication
        .map( a.getAuthorities()
            .contains(new SimpleGrantedAuthority("ROLE_ADMIN")))
```

```
            .map( AuthorizationDecision::new);
}
```

access()表达式可用于编写非常强大的表达式。如果当前用户具有 ROLE_ADMIN 权限，则允许访问上述代码片段。Authentication 实例中包含了 GrantedAuthorities 的集合，可用于检查用户的角色是否是 ROLE_ADMIN。当然，你可以编写任意数量的复杂表达式，用于检查 IP 地址、请求头等。

6.6 小结

在本章中，你学习了如何使用 Spring Security 保护 Spring Boot 应用程序。它可以用来保护任何 Java 应用程序，但主要用于保护基于 Web 的应用程序。身份验证、授权和访问控制是安全领域中的核心概念，因此你应该清晰地理解它们的含义。

通常必须通过防止未经授权的访问来保护关键 URL。Spring Security 可以帮助你以声明的方式实现这一点。它通过应用 Servlet 过滤器来保证应用程序的安全性，过滤器可以用简单的基于 Java 的配置来配置。Spring Security 将自动配置基本的安全服务，并在默认情况下尽量保证安全性。

Spring Security 支持多种用户登录 Web 应用的方式，例如基于表单的登录和 HTTP 基本身份验证。它还提供匿名登录服务，允许你像处理普通用户一样处理匿名用户。"记住我"特性允许应用程序跨多个浏览器会话记住用户的身份。

Spring Security 支持多种方式对用户进行身份验证，并提供了内置的提供商程序实现。例如，它支持根据内存内定义、关系型数据库和 LDAP 存储库对用户进行身份验证。应该始终将加密的密码存储在用户存储库中，因为明文密码很容易受到黑客攻击。Spring Security 还支持本地缓存用户详细信息，以节省执行远程查询的开销。

由访问决策管理器决定是否允许用户访问特定的资源。Spring Security 有三个基于投票机制的访问决策管理器。所有这些访问决策管理器都需要配置一组投票器，以便对访问控制决策进行投票。

Spring Security 允许使用@PreAuthorize 和@PostAuthorize 注解以声明性的方式保护方法调用。

Spring Security 还支持保护基于 Spring WebFlux 的应用程序。在上一小节，本书探讨了如何为这样的应用程序添加安全特性。

第 7 章

数 据 访 问

使用数据库时，首先需要的是与数据库建立连接。在 Java 中与数据库建立连接是通过 javax.sql.DataSource 类实现的。Spring 提供了一些现成的 DataSource 实现，如 DriverManagerDataSource 和 SimpleDriverDataSource。然而，这些实现不是连接池，应该主要考虑用于测试，而不是用于生产。对于商业系统，我们希望使用适当的连接池，比如 HikariCP[1]。

■ 提示：
在db文件夹中有一个名为Dockerfile的文件，它将使用自动创建的数据库构建PostgreSQL[2]。使用命令docker build-t sb2r-postgres构建PostgreSQL，然后可以使用命令docker run-p 5432:5432-it sb2r-postgres运行PostgreSQL。

7.1 配置数据源

7.1.1 问题

你需要从应用程序访问数据库。

7.1.2 解决方案

Spring Boot 通过三个属性 spring.datasource.url、spring.datasource.username 和 spring.datasource.password 配置数据源。

7.1.3 工作原理

为了配置 DataSource，Spring Boot 需要连接池或嵌入式数据库，比如 H2、

---

[1] https://brettwooldridge.github.io/HikariCP/.
[2] https://www.postgresql.org.

HSQLDB 或 Derby。Spring Boot 自动检测基于 HikariCP、Tomcat JDBC 和 Commons DBCP2 的连接池(按此顺序检测)。要使用连接池，需要配置 spring.datasource.url、spring.datasource.username 和 spring.datasource.password 属性。可以通过添加 spring-boot-starter-jdbc 依赖项来启用 JDBC 支持。这将引入使用 JDBC 所需的所有依赖项。

```xml
<dependency>
  <groupId>org.springframework.boot</groupId>
  <artifactId>spring-boot-starter-jdbc</artifactId>
</dependency>
```

上述依赖项引入的依赖项包括 spring-jdbc、spring-tx 和作为默认连接池的 HikariCP。

1. 使用嵌入式 DataSource

当 Spring Boot 检测到存在 H2、HSQLDB 或 Derby 时，默认情况下，它将使用检测到的嵌入式实现启动一个嵌入式数据库。这在编写测试或准备业务演示时非常有用。只需要包含所需的依赖项，Spring Boot 将自动启动相关的数据库。

```xml
<dependency>
  <groupId>org.apache.derby</groupId>
  <artifactId>derby</artifactId>
  <scope>runtime</scope>
</dependency>
```

根据上述代码添加的依赖项，Spring Boot 将检测到 Derby 并启动一个嵌入式的 DataSource。接下来编写一个应用程序，该应用程序列出此数据库中所有的表格。

```java
package com.apress.springboot2recipes.jdbc;

import org.slf4j.Logger;
import org.slf4j.LoggerFactory;
import org.springframework.boot.ApplicationArguments;
import org.springframework.boot.ApplicationRunner;
import org.springframework.boot.SpringApplication;
import org.springframework.boot.autoconfigure.SpringBootApplication;
import org.springframework.stereotype.Component;

import javax.sql.DataSource;
```

第 7 章 ■ 数 据 访 问

```
@SpringBootApplication
public class JdbcApplication {

  public static void main(String[] args) {
      SpringApplication.run(JdbcApplication.class, args);
  }
}

@Component
class TableLister implements ApplicationRunner {

  private final Logger logger = LoggerFactory.getLogger(getClass());
  private final DataSource dataSource;

  TableLister(DataSource dataSource) {
      this.dataSource = dataSource;
  }

  @Override
  public void run(ApplicationArguments args) throws Exception {
      try (var con = dataSource.getConnection();
          var rs = con.getMetaData().getTables(null,null,"%",null)) {
      while (rs.next()) {
        logger.info("{}", rs.getString(3));
      }
    }
  }
}
```

当应用程序运行时，它将创建一个 TableLister 实例，该实例将接收已配置的 DataSource。然后，Spring Boot 将检测到它是一个 ApplicationRunner，并调用 run 方法。run 方法从 DataSource 获取一个 Connection 实例，并使用 DatabaseMetaData 实例(来自 JDBC)来获取数据库中的表。现在，当运行应用程序时，它应该显示类似于图 7-1 所示的输出。

```
2018-09-10 19:27:18.543  WARN 96243 --- [           main] com.zaxxer.hikari.util.DriverDataSource  : Registered driver with driverClass
2018-09-10 19:27:18.946  INFO 96243 --- [           main] com.zaxxer.hikari.pool.PoolBase          : HikariPool-1 - Driver does not sup
2018-09-10 19:27:18.949  INFO 96243 --- [           main] com.zaxxer.hikari.HikariDataSource       : HikariPool-1 - Start completed.
2018-09-10 19:27:19.192  INFO 96243 --- [           main] c.a.springboot2recipes.jdbc.TableLister  : SYSALIASES
2018-09-10 19:27:19.192  INFO 96243 --- [           main] c.a.springboot2recipes.jdbc.TableLister  : SYSCHECKS
2018-09-10 19:27:19.192  INFO 96243 --- [           main] c.a.springboot2recipes.jdbc.TableLister  : SYSCOLPERMS
2018-09-10 19:27:19.192  INFO 96243 --- [           main] c.a.springboot2recipes.jdbc.TableLister  : SYSCOLUMNS
2018-09-10 19:27:19.193  INFO 96243 --- [           main] c.a.springboot2recipes.jdbc.TableLister  : SYSCONGLOMERATES
2018-09-10 19:27:19.193  INFO 96243 --- [           main] c.a.springboot2recipes.jdbc.TableLister  : SYSCONSTRAINTS
2018-09-10 19:27:19.193  INFO 96243 --- [           main] c.a.springboot2recipes.jdbc.TableLister  : SYSDEPENDS
2018-09-10 19:27:19.193  INFO 96243 --- [           main] c.a.springboot2recipes.jdbc.TableLister  : SYSFILES
2018-09-10 19:27:19.196  INFO 96243 --- [           main] c.a.springboot2recipes.jdbc.TableLister  : SYSFOREIGNKEYS
2018-09-10 19:27:19.197  INFO 96243 --- [           main] c.a.springboot2recipes.jdbc.TableLister  : SYSKEYS
2018-09-10 19:27:19.197  INFO 96243 --- [           main] c.a.springboot2recipes.jdbc.TableLister  : SYSPERMS
2018-09-10 19:27:19.197  INFO 96243 --- [           main] c.a.springboot2recipes.jdbc.TableLister  : SYSROLES
2018-09-10 19:27:19.197  INFO 96243 --- [           main] c.a.springboot2recipes.jdbc.TableLister  : SYSROUTINEPERMS
2018-09-10 19:27:19.197  INFO 96243 --- [           main] c.a.springboot2recipes.jdbc.TableLister  : SYSSCHEMAS
2018-09-10 19:27:19.197  INFO 96243 --- [           main] c.a.springboot2recipes.jdbc.TableLister  : SYSSEQUENCES
2018-09-10 19:27:19.197  INFO 96243 --- [           main] c.a.springboot2recipes.jdbc.TableLister  : SYSSTATEMENTS
2018-09-10 19:27:19.197  INFO 96243 --- [           main] c.a.springboot2recipes.jdbc.TableLister  : SYSSTATISTICS
2018-09-10 19:27:19.197  INFO 96243 --- [           main] c.a.springboot2recipes.jdbc.TableLister  : SYSTABLEPERMS
2018-09-10 19:27:19.197  INFO 96243 --- [           main] c.a.springboot2recipes.jdbc.TableLister  : SYSTABLES
2018-09-10 19:27:19.197  INFO 96243 --- [           main] c.a.springboot2recipes.jdbc.TableLister  : SYSTRIGGERS
2018-09-10 19:27:19.198  INFO 96243 --- [           main] c.a.springboot2recipes.jdbc.TableLister  : SYSUSERS
2018-09-10 19:27:19.198  INFO 96243 --- [           main] c.a.springboot2recipes.jdbc.TableLister  : SYSVIEWS
2018-09-10 19:27:19.198  INFO 96243 --- [           main] c.a.springboot2recipes.jdbc.TableLister  : SYSDUMMY1
```

图 7-1　TableLister 应用程序列出的 Derby 数据库表格

2. 使用外部数据库

要连接到数据库，需要一个 JDBC 驱动程序。本节使用 PostgreSQL，所以需要为这个数据库引入该驱动程序。

```
<dependency>
  <groupId>org.postgresql</groupId>
  <artifactId>postgresql</artifactId>
</dependency>
```

现在配置单个数据源只需要在 application.properties 文件中配置相关的属性即可。对于本节使用的 PostgreSQL 数据库，数据源配置如下所示。实际开发中，可以根据需要替换为我们自己的配置。

```
spring.datasource.url=jdbc:postgresql://localhost:5432/customers
spring.datasource.username=customers
spring.datasource.password=customers
```

spring.datasource.url 属性指定数据源连接到哪里，spring.datasource.username 和 spring.datasource.password 属性配置连接时要使用的用户名和密码，还可以配置 spring.datasource.driver-class-name 属性指定要使用的 JDBC 驱动程序类。通常，Spring Boot 会从传入的 URL 检测要使用的驱动程序。如果出于性能或日志记录的原因要使用非默认驱动程序，也可以指定希望使用的驱动程序。

在运行 JDBCApplication 应用程序时，输出应该类似于图 7-2；与 Derby 相比，现在出现了一些不同的表。

```
2018-09-10 19:30:15.711  INFO 97421 --- [           main] com.zaxxer.hikari.HikariDataSource       : HikariPool-1 - Starting...
2018-09-10 19:30:15.996  INFO 97421 --- [           main] com.zaxxer.hikari.HikariDataSource       : HikariPool-1 - Start completed.
2018-09-10 19:30:16.012  INFO 97421 --- [           main] c.a.springboot2recipes.jdbc.TableLister  : pg_aggregate_fnoid_index
2018-09-10 19:30:16.012  INFO 97421 --- [           main] c.a.springboot2recipes.jdbc.TableLister  : pg_am_name_index
2018-09-10 19:30:16.012  INFO 97421 --- [           main] c.a.springboot2recipes.jdbc.TableLister  : pg_am_oid_index
2018-09-10 19:30:16.012  INFO 97421 --- [           main] c.a.springboot2recipes.jdbc.TableLister  : pg_amop_fam_strat_index
2018-09-10 19:30:16.012  INFO 97421 --- [           main] c.a.springboot2recipes.jdbc.TableLister  : pg_amop_oid_index
2018-09-10 19:30:16.012  INFO 97421 --- [           main] c.a.springboot2recipes.jdbc.TableLister  : pg_amop_opr_fam_index
2018-09-10 19:30:16.013  INFO 97421 --- [           main] c.a.springboot2recipes.jdbc.TableLister  : pg_amproc_fam_proc_index
2018-09-10 19:30:16.013  INFO 97421 --- [           main] c.a.springboot2recipes.jdbc.TableLister  : pg_amproc_oid_index
2018-09-10 19:30:16.013  INFO 97421 --- [           main] c.a.springboot2recipes.jdbc.TableLister  : pg_attrdef_adrelid_adnum_index
2018-09-10 19:30:16.013  INFO 97421 --- [           main] c.a.springboot2recipes.jdbc.TableLister  : pg_attrdef_oid_index
2018-09-10 19:30:16.013  INFO 97421 --- [           main] c.a.springboot2recipes.jdbc.TableLister  : pg_attribute_relid_attnam_index
2018-09-10 19:30:16.013  INFO 97421 --- [           main] c.a.springboot2recipes.jdbc.TableLister  : pg_attribute_relid_attnum_index
2018-09-10 19:30:16.013  INFO 97421 --- [           main] c.a.springboot2recipes.jdbc.TableLister  : pg_auth_members_member_role_index
2018-09-10 19:30:16.013  INFO 97421 --- [           main] c.a.springboot2recipes.jdbc.TableLister  : pg_auth_members_role_member_index
2018-09-10 19:30:16.013  INFO 97421 --- [           main] c.a.springboot2recipes.jdbc.TableLister  : pg_authid_oid_index
2018-09-10 19:30:16.013  INFO 97421 --- [           main] c.a.springboot2recipes.jdbc.TableLister  : pg_authid_rolname_index
2018-09-10 19:30:16.013  INFO 97421 --- [           main] c.a.springboot2recipes.jdbc.TableLister  : pg_cast_oid_index
2018-09-10 19:30:16.013  INFO 97421 --- [           main] c.a.springboot2recipes.jdbc.TableLister  : pg_cast_source_target_index
2018-09-10 19:30:16.014  INFO 97421 --- [           main] c.a.springboot2recipes.jdbc.TableLister  : pg_class_oid_index
2018-09-10 19:30:16.014  INFO 97421 --- [           main] c.a.springboot2recipes.jdbc.TableLister  : pg_class_relname_nsp_index
2018-09-10 19:30:16.014  INFO 97421 --- [           main] c.a.springboot2recipes.jdbc.TableLister  : pg_class_tblspc_relfilenode_index
2018-09-10 19:30:16.014  INFO 97421 --- [           main] c.a.springboot2recipes.jdbc.TableLister  : pg_collation_name_enc_nsp_index
2018-09-10 19:30:16.014  INFO 97421 --- [           main] c.a.springboot2recipes.jdbc.TableLister  : pg_collation_oid_index
```

图 7-2　TableLister 应用程序列出的 PostgreSQL 数据库表格

3．从 JNDI 获取 DataSource

如果你正在将 Spring Boot 应用程序部署到应用服务器(或者，如果你有远程 JNDI 服务器)，并且想要使用预配置的 DataSource，那么可以使用 spring.datasource.jndi-name 属性让 Spring Boot 知道你想要从 JNDI 获取 DataSource。

```
spring.datasource.jndi-name=java:jdbc/customers
```

4．配置连接池

Spring Boot 使用的默认连接池是 HikariCP。当引入 spring-boot-starter-jdbc 依赖项(或与其他数据库相关的依赖项之一)时，Spring Boot 会自动完成相关配置。Spring Boot 将使用一些默认设置来配置连接池，但你可以根据需要覆盖这些设置，例如增加或减少最大连接数、设置超时时间等。HikariCP 的配置选项位于 spring.datasource.hikari 名称空间中，如表 7-1 所示。

表 7-1　HikariCP 通用连接池设置

属性	描述
spring.datasource.hikari.connection-timeout	客户端等待连接池分配连接的最大毫秒数。默认值为 30 秒
spring.datasource.hikari.leak-detection-threshold	在日志中记录相关消息以指示可能的连接泄漏之前，可以从池中取出一个连接的毫秒数。默认值为 0(等同于不允许)
spring.datasource.hikari.idle-timeout	允许连接在连接池中空闲的最大毫秒数。默认值为 10 分钟
spring.datasource.hikari.validation-timeout	连接池等待连接验证为活跃连接的最大毫秒数。默认值为 5 秒

(续表)

属性	描述
spring.datasource.hikari.connection-test-query	要执行的用于测试连接有效性的 SQL 查询。注意：通常 JDBC 4.0(或更高版本的)驱动程序不需要进行测试！默认为空
spring.datasource.hikari.maximum-pool-size	连接池能保留的最大连接数。默认为 10 个连接
spring.datasource.hikari.minimum-idle	池中维护的最少空闲连接数。默认为 10 个连接

还有很多属性可以使用，但全部列出来表格会很长，表 7-1 仅列出了最常用的属性。

■ 提示：
Tomcat JDBC的属性位于spring.datasource.tomcat名称空间中，Commons DBCP2的属性位于spring.datasource.dbcp2 名称空间中。

要将数据源配置为最多 5 个连接、最少两个连接和泄漏检测阈值为 20 秒，需要添加以下配置。

```
spring.datasource.hikari.maximum-pool-size=5
spring.datasource.hikari.minimum-idle=2
spring.datasource.hikari.leak-detection-threshold=20000
```

5. 使用 Spring Boot 初始化数据库

使用现有数据库时，数据库中可能已经存在现有的表、视图和存储过程。但是，在创建一个新数据库时，它是空的，你需要自己创建表。Spring Boot 为此提供现成的支持。可以添加 schema.sql 文件来初始化模式(即表、视图等)，添加 data.sql 文件将数据插入表中。Spring Boot 还允许你提供 schema-<db-platform>.sql 和 data-<db-platform>.sql 文件以执行特定于数据库的初始化。<db-platform>的值从 spring.datasource.platform 属性中读取(见表 7-2)。例如，在使用 Derby 时，可以添加 schema-derby.sql 文件，以此类推。schema-<db-platform>.sql 和 data-<db-platform>.sql 文件的名称可以通过 spring.datasource.schema 和 spring.datasource.data 属性更改。可用属性说明见表 7-2。

表 7-2　数据源初始化属性

属性	说明
spring.datasource.continue-on-error	初始化数据库时如果出现错误，是否停止，默认为 false
spring.datasource.data	数据(DML)脚本资源引用，默认为 classpath:data
spring.datasource.data-password	执行 DML 脚本的数据库密码，默认为普通密码

(续表)

属性	说明
spring.datasource.data-username	执行 DML 脚本的数据库用户名,默认为普通用户名
spring.datasource.initialization-mode	使用可用的 DDL 和 DML 脚本初始化数据源。默认为 embedded,可以更改为 never 或 always
spring.datasource.platform	在 DDL 或 DML 脚本(例如 schema-${platform.sql} 或 data-${platform}.sql)中使用的平台。默认为 all
spring.datasource.schema	数据(DDL)脚本资源引用,默认为 classpath:schema
spring.datasource.schema-password	执行 DDL 脚本的数据库密码,默认为普通密码
spring.datasource.schema-username	执行 DDL 脚本的数据库用户名,默认为普通用户名
spring.datasource.separator	SQL 初始化脚本中语句的分隔符,默认为分号";"
spring.datasource.sql-script-encoding	SQL 脚本编码,默认为平台编码

让我们创建 customer 表并在其中插入一些数据。要创建表,请在 src/main/resources 目录中添加 schema.sql 文件,并将如下语句放入该文件中:

```
DROP TABLE IF EXISTS customer;

CREATE TABLE customer (
  id SERIAL PRIMARY KEY,
  name VARCHAR(100) NOT NULL,
  email VARCHAR(255) NOT NULL,
  UNIQUE(name)
);
```

要插入数据,请在 src/main/resources 目录下添加 data.sql 文件,并将如下语句放入该文件中:

```
INSERT INTO customer (name, email) VALUES
  ('Marten Deinum', 'marten.deinum@conspect.nl'),
  ('Josh Long', 'jlong@pivotal.com'),
  ('John Doe', 'john.doe@island.io'),
  ('Jane Doe', 'jane.doe@island.io');
```

为了查看插入数据是否成功,让我们添加另一个 ApplicationRunner,它使用 DataSource 打印 customer 表的内容。

```
package com.apress.springboot2recipes.jdbc;

// Imports removed
```

```java
@SpringBootApplication
public class JdbcApplication {

   public static void main(String[] args) {
       SpringApplication.run(JdbcApplication.class, args);
   }
}

... // Other code remove
@Component
class CustomerLister implements ApplicationRunner {

  private final Logger logger = LoggerFactory.getLogger(getClass());
  private final DataSource dataSource;

  CustomerLister(DataSource dataSource) {
     this.dataSource = dataSource;
  }

  @Override
  public void run(ApplicationArguments args) throws Exception {
     var query = "SELECT id, name, email FROM customer";
     try (var con = dataSource.getConnection();
        var stmt = con.createStatement();
        var rs = stmt.executeQuery(query)) {
        while (rs.next()) {
          logger.info("Customer [id={}, name={}, email={}]",
          rs.getLong(1), rs.getString(2), rs.getString(3));
        }
     }
  }
}
```

对于嵌入式数据库，数据库初始化功能始终处于开启状态，因此当使用 Derby、H2 或 HSQLDB 时，初始化功能是默认打开的。使用外部数据库时，默认情况下不会进行初始化。要更改此设置，可以将 spring.datasource.initialization-mode 属性切换为 always，以便初始化会始终运行。

```
spring.datasource.initialization-mode=always
```

当应用程序启动时，你将在日志中看到正在打印的客户清单。

6. 使用 Flyway 初始化数据库

在开发应用程序时，你需要对数据库迁移进行更多的控制。使用 schema.sql 和 data.sql 可以稳定而迅速地完成迁移，但是脚本最终维护起来会很麻烦。Spring Boot 还支持 Flyway[3]，简单地说，它是数据库模式的版本控制工具。它允许你增量地更改/更新数据库模式。要使用 Flyway，首先要做的是添加 Flyway 本身的依赖项。

```xml
<dependency>
    <groupId>org.flywaydb</groupId>
    <artifactId>flyway-core</artifactId>
</dependency>
```

Spring Boot 将检测到 Flyway 的存在，并假设你希望使用它来进行数据库迁移。迁移脚本应该位于 src/main/resources 中的 db/migration 文件夹中。

```sql
CREATE TABLE customer (
  id SERIAL PRIMARY KEY,
  name VARCHAR(100) NOT NULL,
  email VARCHAR(255) NOT NULL,
  UNIQUE(name)
);
INSERT INTO customer (name, email) VALUES
  ('Marten Deinum', 'marten.deinum@conspect.nl'),
  ('Josh Long', 'jlong@pivotal.com'),
  ('John Doe', 'john.doe@island.io'),
  ('Jane Doe', 'jane.doe@island.io');
```

当将上述 SQL 语句放入 db/migration 文件夹中的 V1__first.sql 文件中时，应用程序启动时将执行此脚本文件(假定数据库为空)。默认情况下，命名约定是 v<sequence>__<name>.sql，用于确定要执行的操作。一旦执行了脚本，就不能(也不应该)修改该脚本，否则 Flyway 将阻止应用程序启动，因为它检测到已经执行的脚本被修改了。

在运行应用程序时，日志中仍将打印客户清单。请注意表格清单(如图 7-3 所示)中增加了一个表，即 flyway_schema_history。此表包含 Flyway 用于检测(和保护)数据库变更的元数据。

---

[3] https://flywaydb.org.

```
2018-09-10 19:57:22.714  INFO 98933 --- [           main] c.a.springboot2recipes.jdbc.TableLister  : pg_user
2018-09-10 19:57:22.714  INFO 98933 --- [           main] c.a.springboot2recipes.jdbc.TableLister  : pg_user_mappings
2018-09-10 19:57:22.714  INFO 98933 --- [           main] c.a.springboot2recipes.jdbc.TableLister  : pg_views
2018-09-10 19:57:22.714  INFO 98933 --- [           main] c.a.springboot2recipes.jdbc.TableLister  : customer
2018-09-10 19:57:22.714  INFO 98933 --- [           main] c.a.springboot2recipes.jdbc.TableLister  : flyway_schema_history
2018-09-10 19:57:22.714  INFO 98933 --- [           main] c.a.springboot2recipes.jdbc.TableLister  : pg_toast_12242
2018-09-10 19:57:22.714  INFO 98933 --- [           main] c.a.springboot2recipes.jdbc.TableLister  : pg_toast_12247
2018-09-10 19:57:22.715  INFO 98933 --- [           main] c.a.springboot2recipes.jdbc.TableLister  : pg_toast_12252
```

图 7-3　使用 Flyway 时的表格清单

可以通过 Spring Boot 提供的一些属性来配置 Flyway，最常用的属性见表 7-3。

表 7-3　常用的 Flyway 配置属性

属性	说明
spring.flyway.enabled	是否启用 Flyway，默认为 true
spring.flyway.locations	迁移脚本的位置，默认为 classpath:db/migration
spring.flyway.url	待迁移数据库的 JDBC URL，未设置时则使用默认配置的 DataSource
spring.flyway.user	如果 Flyway 使用自己的 DataSource 配置，则指定用于访问数据库的用户名
spring.flyway.password	如果 Flyway 使用自己的 DataSource 配置，则指定用于访问数据库的用户密码

7.2　使用 JdbcTemplate

7.2.1　问题

你希望使用 JdbcTemplate 或 NamedParameterJdbcTemplate 以便更方便地使用 JDBC。

7.2.2　解决方案

使用自动配置的 JdbcTemplate 或 NamedParameterJdbcTemplate 来执行查询并处理结果。

7.2.3　工作原理

默认情况下，Spring Boot 在检测到单个候选 DataSource 时将配置 JdbcTemplate 和 NamedParameterjdbcTemplate。单个候选 DataSource 意味着只有一个 DataSource，或者有一个使用@Primary 标记为主资源的 DataSource。由于 JdbcTemplate 已配置好，因此可以使用它编写 JDBC 代码。使用 JdbcTemplate 重写 CustomerLister，替换原来的普通 DataSource，将使代码更容易阅读，从而更容易处理。

■ 注意：
NamedParameterJdbcTemplate与JdbcTemplate类似，其主要优点是能够在查询中使用命名参数，而不是使用常规的JDBC占位符。

```
@Component
class CustomerLister implements ApplicationRunner {

  private final Logger logger = LoggerFactory.getLogger(getClass());
  private final JdbcTemplate jdbc;

  CustomerLister(JdbcTemplate jdbc) {
     this.jdbc = jdbc;
  }

  @Override
  public void run(ApplicationArguments args) {

     var query = "SELECT id, name, email FROM customer";
     jdbc.query(query, rs -> {
        logger.info("Customer [id={}, name={}, email={}]",
          rs.getLong(1), rs.getString(2), rs.getString(3));
        });
     }
}
```

JdbcTemplate 用于通过 query 方法执行查询，此方法接收两个参数，第一个是字符串，第二个是 RowCallbackHandler 实例。JdbcTemplate 将执行查询，并且对于每一行数据调用一次 RowCallbackHandler，它将该行数据记录到日志中。运行应用程序时，输出不变，但代码变得更清晰。

JdbcTemplate 还有一个让我们觉得更熟悉的接口，即 RowMapper 接口，它可以用来将表格中的一行数据从 ResultSet 映射为一个 Java 对象。让我们创建 Customer 类，并使用 RowMapper 接口从数据库创建 Customer 实例。

```
package com.apress.springboot2recipes.jdbc;

import java.util.Objects;

public class Customer {
```

```java
    private final long id;
    private final String name;
    private final String email;

    Customer(long id, String name, String email) {
      this.id = id;
      this.name = name;
      this.email = email;
    }

    public long getId() {
      return id;
    }

    public String getName() {
        return name;
    }

    public String getEmail() {
       return email;
    }

    @Override
    public boolean equals(Object o) {
        if (this == o) return true;
        if (o == null || getClass() != o.getClass()) return false;
        Customer customer = (Customer) o;
        return id == customer.id &&
          Objects.equals(name, customer.name) &&
          Objects.equals(email, customer.email);
    }

    @Override
    public int hashCode() {
        return Objects.hash(id, name, email);
    }
```

第 7 章 ■ 数 据 访 问

```java
    @Override
    public String toString() {
      return "Customer [" +
         "id=" + id +", name='" + name + '\"' +
         ", email='" + email + '\"' + ']';
      }
  }
```

接下来，我们创建一个存储库接口来定义交换数据的格式，并创建一个基于 JDBC 的实现。

```java
package com.apress.springboot2recipes.jdbc;

import java.util.List;

public interface CustomerRepository {
   List<Customer> findAll();
   Customer findById(long id);
   Customer save(Customer customer);
}
```

接下来，代码使用 JdbcTemplate 和 RowMapper 将结果映射到 Customer 对象。

```java
package com.apress.springboot2recipes.jdbc;

import org.springframework.jdbc.core.JdbcTemplate;
import org.springframework.jdbc.support.GeneratedKeyHolder;
import org.springframework.stereotype.Repository;

import java.sql.PreparedStatement;
import java.sql.ResultSet;
import java.sql.SQLException;
import java.util.List;

@Repository
class JdbcCustomerRepository implements CustomerRepository {

  private static final String ALL_QUERY =
      "SELECT id, name, email FROM customer";
  private static final String BY_ID_QUERY =
```

```java
            "SELECT id, name, email FROM customer WHERE id=?";
    private static final String INSERT_QUERY =
            "INSERT INTO customer (name, email) VALUES (?,?)";
    private final JdbcTemplate jdbc;

    JdbcCustomerRepository(JdbcTemplate jdbc) {
        this.jdbc = jdbc;
    }

    @Override
    public List<Customer> findAll() {
        return jdbc.query(ALL_QUERY, (rs, rowNum) -> toCustomer(rs));
    }

    @Override
    public Customer findById(long id) {
        return jdbc.queryForObject(BY_ID_QUERY, (rs, rowNum) ->
            toCustomer(rs), id);
    }

    @Override
    public Customer save(Customer customer) {
        var keyHolder = new GeneratedKeyHolder();
        jdbc.update(con -> {
          var ps = con.prepareStatement(INSERT_QUERY);
          ps.setString(1, customer.getName());
          ps.setString(2, customer.getEmail());
          return ps;
        }, keyHolder);
        return new Customer(keyHolder.getKey().longValue(),
                            customer.getName(), customer.getEmail());
    }

    private Customer toCustomer(ResultSet rs) throws SQLException {
      var id = rs.getLong(1);
      var name = rs.getString(2);
      var email = rs.getString(3);
      return new Customer(id, name, email);
```

                }
            }

如上所示，JdbcCustomerRepository 类使用 JdbcTemplate 和 RowMapper(通过 lambda 表达式)将 ResultSet 转换到 Customer 实例中。CustomerLister 现在可以使用 CustomerRepository 从数据库中获取所有 Customer 数据，并将它们打印到控制台。

```java
@Component
class CustomerLister implements ApplicationRunner {

    private final Logger logger = LoggerFactory.getLogger(getClass());
    private final CustomerRepository customers;
    CustomerLister(CustomerRepository customers) {
        this.customers = customers;
    }

    @Override
    public void run(ApplicationArguments args) {

        customers.findAll()
            .forEach( customer -> logger.info("{}", customer));
    }
}
```

由于所有 JDBC 代码都已转移到 JDBCCustomerRepository 类中，因此这个类变得非常简单。它会接收一个注入的 CustomerRepository 实例，并使用该实例的 findAll 方法获取数据库的内容，然后为每个客户打印一行信息。

测试 JDBC 代码

测试 JDBC 代码需要数据库，通常使用嵌入式数据库(如 H2、Derby 或 HSQLDB)进行测试。Spring Boot 使得为 JDBC 代码编写测试非常容易。基于 JDBC 的测试可以用@JdbcTest 注解，Spring Boot 将创建一个最小化的应用程序，只使用与 JDBC 相关的 bean，如 DataSource 和事务管理器。

让我们为 JdbcCustomerRepository 编写一个测试，使用的是嵌入式数据库 H2。首先，将 H2 作为测试依赖项添加到依赖项清单中。

```xml
<dependency>
    <groupId>com.h2database</groupId>
    <artifactId>h2</artifactId>
    <scope>test</scope>
```

```
</dependency>
```

接下来创建 JdbcCustomerRepositoryTest 类。

```
@RunWith(SpringRunner.class)
@JdbcTest(includeFilters =
  @ComponentScan.Filter(
    type= FilterType.REGEX,
    pattern = "com.apress.springboot2recipes.jdbc.*Repository"))
@TestPropertySource(properties = "spring.flyway.enabled=false")
public class JdbcCustomerRepositoryTest {

  @Autowired
  private JdbcCustomerRepository repository;
}
```

@RunWith(SpringRunner.class)通过一个特殊的 JUnit 运行程序执行测试，从而启动 Spring 测试上下文框架。@JdbcTest 将用一个嵌入式数据源(在本例中为 H2)替换预配置的 DataSource。由于我们还想创建存储库的实例，因此在@JdbcTest 中添加了 includeFilters 属性并为其提供一个正则表达式来匹配我们的 JdbcCustomerRepository。

JdbcCustomerRepositoryTest 类的最后一个注解是 @TestPropertySource (properties="spring.flyway.enabled=false")，它表明我们要禁用 Flyway。应用程序使用 Flyway 管理模式，但是，这些脚本是为 PostgreSQL 而不是 H2 编写的。对于测试，我们希望禁用 Flyway，并提供一个基于 H2 的 schema.sql 文件来创建模式。

■ 注意：
这是使用不同于实际运行系统的数据库进行测试的缺点之一，实际系统使用 PostgreSQL 数据库，而测试使用了 H2 数据库。你要么需要为模式维护两组脚本，要么使用与实时系统相同的数据库进行测试。

在 src/test/resources 目录下添加 schema.sql 文件，并将如下 DDL 语句放入该文件：

```
CREATE TABLE customer (
  id BIGINT AUTO_INCREMENT PRIMARY KEY,
  name VARCHAR(100) NOT NULL,
  email VARCHAR(255) NOT NULL,
  UNIQUE(name)
);
```

现在编写测试，测试数据是否正确地插入表格中。

```
@Test
public void insertNewCustomer() {
  assertThat(repository.findAll()).isEmpty();

  Customer customer = repository.save(new Customer(-1, "T. Testing",
  "t.testing@test123.tst"));

  assertThat(customer.getId()).isGreaterThan(-1L);
  assertThat(customer.getName()).isEqualTo("T. Testing");
  assertThat(customer.getEmail()).isEqualTo("t.testing
  @test123.tst");

  assertThat(repository.findById(customer.getId())).isEqualTo
  (customer);
}
```

该测试首先断言数据库是空的——不一定需要这样做，但可以检测其他测试是否污染了数据库。接下来，通过调用 JdbcCustomerRepository 类的 save 方法将 Customer 添加到数据库中。对插入了数据的 Customer 表格，使用断言验证其 id 字段不为空，以及 name 和 email 属性的值已保存在表格中。最后，再次从 Customer 表格检索数据并与插入前的数据进行比较。

另一个可以测试的是 findAll 方法。当插入两条记录时，调用 findAll 方法将检索两条记录。

```
@Test
public void findAllCustomers() {
  assertThat(repository.findAll()).isEmpty();

  repository.save(new Customer(-1,"T. Testing1",
    "t.testing@test123.tst"));
  repository.save(new Customer(-1, "T. Testing2",
    "t.testing@test123.tst"));

  assertThat(repository.findAll()).hasSize(2);
}
```

当然可以使用更多断言进行其他测试，但数据保存的问题已经在其他测试方法中验证过了。

7.3 使用 JPA

7.3.1 问题

你希望在 Spring Boot 应用程序中使用 JPA。

7.3.2 解决方案

Spring Boot 会自动检测到 Hibernate，所需的 JPA 类将使用这些信息来配置一个 EntityManagerFactory 实例。

7.3.3 工作原理

Spring Boot 通过 Hibernate 对 JPA 提供开箱即用的支持[4]。当检测到 Hibernate 时，将使用前面配置的 DataSource 自动配置一个 EntityManagerFactory 实例(参见 7.1 小节)。

首先，需要将 hibernate-core 和 spring-orm 作为依赖项添加到项目中。但是，将 spring-boot-starter-data-jpa 依赖项添加到项目中会更容易一些(尽管这也会引入 spring-data-jpa 作为一个依赖项)。

```xml
<dependency>
    <groupId>org.springframework.boot</groupId>
    <artifactId>spring-boot-starter-data-jpa</artifactId>
</dependency>
```

这将为类路径添加所有必需的依赖项。

1. 直接使用 JPA 存储库

为了使用 JPA，必须对系统中作为实体的类进行注解。在系统中，我们将 Customer 存储到数据库中并从数据库检索出来。我们需要使用@Entity 注解将其标记为一个实体。JPA 实体类必须有一个默认的无参数构造函数(尽管它可以是包的私有函数)和一个用@id 标记的字段作为主键。

```
@Entity
public class Customer {

    @Id
```

---

[4] https://www.hibernate.org.

```
    @GeneratedValue(strategy = GenerationType.IDENTITY)
    private long id;

    @Column(nullable = false)
    private final String name;

    @Column(nullable = false)
    private final String email;

    Customer() {
        this(null,null);
    }
    // Other code omitted
}
```

接下来，创建 CustomerRepository 类的 JPA 实现(参见 7.2 小节)。要使用 JPA，必须获取 EntityManager 实例。为此，需要声明一个字段并使用@PersistenceContext 对其进行注解。

```
package com.apress.springboot2recipes.jpa;

import org.springframework.stereotype.Repository;

import javax.persistence.EntityManager;
import javax.persistence.PersistenceContext;
import java.util.List;

@Repository
class JpaCustomerRepository implements CustomerRepository {

    @PersistenceContext
    private EntityManager em;

    @Override
    public List<Customer> findAll() {
        var query = em.createQuery("SELECT c FROM Customer c",
            Customer.class);
        return query.getResultList();
    }
```

```java
    @Override
    public Customer findById(long id) {
        return em.find(Customer.class, id);
    }

    @Override
    public Customer save(Customer customer) {
        em.persist(customer);
        return customer;
    }
}
```

下面的应用程序类(与 7.2 小节中的类相似)将从数据库中读取所有 Customer 数据并将它们打印到日志中，如图 7-4 所示。

```java
package com.apress.springboot2recipes.jpa;

import org.slf4j.Logger;
import org.slf4j.LoggerFactory;
import org.springframework.boot.ApplicationArguments;
import org.springframework.boot.ApplicationRunner;
import org.springframework.boot.SpringApplication;
import org.springframework.boot.autoconfigure.SpringBootApplication;
import org.springframework.stereotype.Component;

@SpringBootApplication
public class JpaApplication {

  public static void main(String[] args) {
      SpringApplication.run(JpaApplication.class, args);
  }
}

@Component
class CustomerLister implements ApplicationRunner {
  private final Logger logger = LoggerFactory.getLogger(getClass());
  private final CustomerRepository customers;
```

```
    CustomerLister(CustomerRepository customers) {
        this.customers = customers;
    }

    @Override
    public void run(ApplicationArguments args) {

        customers.findAll()
            .forEach( customer -> logger.info("{}", customer));
    }
}
```

图 7-4　JPA 打印的 Customer 信息

可以使用一些配置选项来配置应用程序中的 EntityManagerFactory 类，这些属性位于 spring.jpa 名称空间中，如表 7-4 所示。

表 7-4　JPA 属性

属性	描述
spring.jpa.database	要操作的目标数据库，默认情况下自动检测
spring.jpa.database-platform	要操作的目标数据库的名称，默认情况下自动检测。可用于指定要使用的特定 Hibernate Dialect
spring.jpa.generate-ddl	启动时初始化架构，默认值为 false
spring.jpa.show-sql	启用 SQL 语句的日志记录，默认值为 false
spring.jpa.open-in-view	注册 OpenEntityManagerInViewInterceptor。将 EntityManager 绑定到请求处理线程。默认为 true
spring.jpa.hibernate.ddl-auto	hibernate.hbm2ddl.auto 属性的简写，默认为 none。如果是嵌入式数据库，可以设置为 create-drop，表示每次应用程序结束时将清空表
spring.jpa.hibernate.use-new-id-generator-mappings	hibernate.id.new_generator_mappings 属性的简写。未显式设置时，默认值为 true
spring.jpa.hibernate.naming.implicit-strategy	隐式命名策略的 FQN(Fully Qualified Name，完全限定名称)，默认为 org.springframework.boot.orm.jpa.hibernate.SpringPhysicalNamingStrategy

(续表)

属性	描述
spring.jpa.hibernate.naming.physical-strategy	物理命名策略的 FQN，默认的 org.springframework.boot.orm.jpa.hibernate.SpringImplicitNamingStrategy
spring.jpa.mapping-resources	附加的 XML 文件，包含了用 XML 而非 Java 描述的实体映射信息。类似于 JPA 特有的 orm.xml 文件
spring.jpa.properties.*	在 JPA 提供商程序上设置的其他属性

如果希望配置 Hibernate 的高级特性，例如设置获取数据大小的属性 hibernate.jdbc.fetch_size 或设置批处理大小的属性 hibernate.jdbc.batch_size，就需要用到 spring.jpa.properties.属性。

```
spring.jpa.properties.hibernate.jdbc.fetch_size=250
spring.jpa.properties.hibernate.jdbc.batch_size=50
```

这将改变 JPA 提供商程序的属性。

2. 使用 Spring Data JPA 存储库

除了编写自己的存储库(这可能是一项单调而重复的任务)以外，也可以让 Spring Data JPA[5]为你完成多种繁重的任务。不需要编写自己的实现，直接扩展 Spring Data 的 CrudRepository 接口并使用 JPA 提供的运行时存储库，这样可以避免编写数据访问代码。当 Spring Boot 在类路径上检测到 Spring Data JPA 时，它还将自动配置 Spring Data JPA。

```
public interface CustomerRepository extends CrudRepository<Customer, Long> { }
```

这就是所需要的一切。Spring Data JPA 提供了 findAll、findById 和 save 等方法。现在可以删除本书前面提供的 JpaCustomerRepository 实现。由于实现接口的形式是 CrudRepository<Customer, Long>，Spring Data 知道它可以查询 Customer 实例，并且它有一个 Long 类型的 id 字段。

运行应用程序得到的输出应该和前面相同(参见图 7-4)。

只有一个属性可以用于配置 Spring Data JPA，那就是显式地启用或禁用它，默认情况下它是启用状态。将 spring.data.jpa.repositories.enabled 设置为 false 将禁用 Spring Data JPA。

3. 使用不同包中的实体

默认情况下，Spring Boot 将从@SpringBootApplication 注解类所在的包开始检

---

5 https://projects.spring.io/spring-data-jpa/.

测组件、存储库和实体。但是，如果实体在另一个包中，却仍然需要包含这些实体呢？为此，可以使用@EntityScan 注解，它的工作方式与@ComponentScan 类似，但只作用于@Entity 注解的 bean。

```
package com.apress.springboot2recipes.order;

import javax.persistence.Entity;
import javax.persistence.Id;
import java.util.Objects;

@Entity
public class Order {

@Id
  private long id;
  private String number;

  public long getId() {
      return id;
  }

  public void setId(long id) {
      this.id = id;
  }

  public String getNumber() {
      return number;
  }

  public void setNumber(String number) {
      this.number = number;
  }
  // Other methods omitted
}
```

这个 Order 类位于 com.apress.springboot2recipes.order 包中，而@SpringBootApplication 注解的类并不覆盖这个类，因为它在 com.apress.springboot2.recipes.jpa 包中。要检测到这个实体，可以将 @EntityScan 注解与需要扫描的包一起添加到由@SpringBootApplication 注解的类(或由@Configuration 注解的常规类)中，如以下代

码所示。

```
@SpringBootApplication
@EntityScan({
  "com.apress.springboot2recipes.order",
  "com.apress.springboot2recipes.jpa" })
```

添加了@EntityScan 注解以及需要扫描的包之后，JPA 将能检查并访问这个 Order 类。

4. 测试 JPA 存储库

在测试 JPA 代码时，需要用到数据库，通常使用嵌入式数据库，例如 H2、Derby 或 HSQLDB。Spring Boot 使得为 JPA 编写测试变得非常容易。基于 JPA 的测试可以使用@DataJpaTest 进行注解，Spring Boot 将创建一个最小化的应用程序，只包含与 JPA 相关的 beans，比如 DataSource、事务管理器，以及 Spring Data JPA 存储库，如果需要的话。

让我们为 CustomerRepository 编写测试并使用 H2 作为嵌入式数据库。首先，将 H2 添加为测试依赖项。

```xml
<dependency>
  <groupId>com.h2database</groupId>
  <artifactId>h2</artifactId>
  <scope>test</scope>
</dependency>
```

接下来，创建 CustomerRepositoryTest 类。

```java
@RunWith(SpringRunner.class)
@DataJpaTest
@TestPropertySource(properties = "spring.flyway.enabled=false")
public class CustomerRepositoryTest {

  @Autowired
  private CustomerRepository repository;

  @Autowired
  private TestEntityManager testEntityManager
}
```

@RunWith(SpringRunner.class)通过一个特殊的 JUnit 运行器执行测试，以启动 Spring Test Context 框架。@DataJpaTest 将用一个嵌入式数据源(在本例中为 H2)替换

预配置的 DataSource。它还将启动 JPA 组件，并在检测到 Spring Data JPA 存储库时启动该存储库。

Spring Boot 还提供了一个 TestEntityManager 实例，它提供了一些方法来轻松地存储和查找用于测试的数据。

最后是@TestPropertySource(properties="spring.flyway.enabled=false")，表示要禁用 Flyway。应用程序使用 Flyway 来管理模式，但是，这些脚本是为 PostgreSQL 而不是 H2 编写的。对于测试，可以让 Hibernate 管理模式，这也是在 Spring Boot 使用嵌入式数据库时的默认设置。

现在，编写一个测试来检查是否正确地插入了记录。

```
@Test
public void insertNewCustomer() {
  assertThat(repository.findAll()).isEmpty();

  Customer customer = repository.save(new Customer(-1, "T. Testing",
  "t.testing@test123.tst"));

  assertThat(customer.getId()).isGreaterThan(-1L);
  assertThat(customer.getName()).isEqualTo("T. Testing");
  assertThat(customer.getEmail()).isEqualTo
    ("t.testing@test123.tst");

  assertThat(repository.findById(customer.getId())).isEqualTo
    (customer);
}
```

该测试首先断言数据库是空的——不一定需要这样做，但可以检测其他测试是否污染了数据库。接下来，通过调用 CustomerRepository 类的 save 方法将一个 Customer 对象插入到数据库中。对插入了数据的 Customer 表格，使用断言验证其 id 字段不为空，以及 name 和 email 属性的值已保存在表格中。最后，再次从 Customer 表格检索数据并与插入前的数据进行比较。

另一个可以测试的是 findAll 方法。当插入两条记录时，调用 findAll 方法将检索两条记录。

```
@Test
public void findAllCustomers() {
  assertThat(repository.findAll()).isEmpty();

  repository.save(new Customer(-1, "T. Testing1",
    "t.testing@test123.tst"));
```

```
repository.save(new Customer(-1, "T. Testing2",
  "t.testing@test123.tst"));

assertThat(repository.findAll()).hasSize(2);
}
```

当然可以使用更多断言进行其他测试，但数据保存的问题已经在其他测试方法中进行了验证。

7.4 直接使用 Hibernate

7.4.1 问题

希望把一些直接使用了 Hibernate API、Session 和/或 SessionFactory 的代码放到 Spring Boot 中。

7.4.2 解决方案

使用 EntityManager 或 EntityManagerFactory 直接获取 Hibernate 的对象，比如 Session 或 SessionFactory。

7.4.3 工作原理

当将旧代码迁移到 Spring Boot 或使用第三方库时，将所有内容迁移到 JPA 可能并不总是可行的。通常你希望按原样使用该代码，或者尽可能少地做修改。有两种方法可以直接使用 Hibernate API：通过从当前的 EntityManager 获取 Session，或者对 SessionFactory 进行配置。具体选择使用哪种方法取决于直接使用 Hibernate API 对代码的影响。

1. 使用 EntityManager 获取 Session

要获取底层的 Hibernate Session 对象，可以使用 unwrap 方法。JPA 2.0 添加了这个方法，当你真正需要访问本地类时，该方法提供一种统一且方便的方式来获取底层的 JPA 实现者类。

```
package com.apress.springboot2recipes.jpa;

import org.hibernate.Session;
import org.springframework.stereotype.Repository;

import javax.persistence.EntityManager;
import javax.persistence.PersistenceContext;
```

```java
import javax.transaction.Transactional;
import java.util.List;
@Repository
@Transactional
class HibernateCustomerRepository implements CustomerRepository {

    @PersistenceContext
    private EntityManager em;

    private Session getSession() {
        return em.unwrap(Session.class);
    }

    @Override
    public List<Customer> findAll() {
        return getSession().createQuery("SELECT c FROM Customer c",
            Customer.class).getResultList();
    }

    @Override
    public Customer findById(long id) {
        return getSession().find(Customer.class, id);
    }

    @Override
    public Customer save(Customer customer) {
        getSession().persist(customer);
        return customer;
    }
}
```

HibernateCustomerRepository 类获取注入的 EntityManager，并使用 unwrap 方法获取 Session 实例(参见 getSession 方法)。现在可以使用 Hibernate Session 来执行查询等操作。

2. 使用 SessionFactory

从 Hibernate 5.2 开始，Hibernate SessionFactory 扩展了 EntityManagerFactory，因此可以用于创建 Session 和 EntityManager。Spring 5.1 在 LocalSessionFactoryBean 中添加了对上述功能的支持，可以将原生 Hibernate 代码和 JPA 代码混合使用。

要配置一个 SessionFactory bean，可以使用 LocalSessionFactoryBean 实例，如以下代码所示。

```
@Bean
public LocalSessionFactoryBean sessionFactory(DataSource dataSource) {

  Properties properties = new Properties();
  properties.setProperty(DIALECT,
      "org.hibernate.dialect.PostgreSQL95Dialect");

  LocalSessionFactoryBean sessionFactoryBean = new
  LocalSessionFactoryBean();
  sessionFactoryBean.setDataSource(dataSource);
  sessionFactoryBean.setPackagesToScan
  ("com.apress.springboot2recipes.jpa");
  sessionFactoryBean.setHibernateProperties(properties);
  return sessionFactoryBean;
}
```

SessionFactory 需要一个 DataSource 以便建立与数据库的连接，需要一种方言来创建 SQL，并且需要知道实体在哪里。这是 SessionFactory 所需的最小配置。不需要添加 HibernateTransactionManager，因为默认配置的 JpaTransactionManager 可以处理事务。然而，由于 JPA 有一些限制(特别是在事务传播级别上的限制)，你可能希望添加 HibernateTransactionManager，它不会受到这些限制的影响。

修改后的 HibernateCustomerRepository 使用 SessionFactory 中的 getCurrentSession 方法获取 Session 实例。

```
@Repository
@Transactional
class HibernateCustomerRepository implements CustomerRepository {

  private final SessionFactory sf;

  HibernateCustomerRepository(SessionFactory sf) {
      this.sf=sf;
  }

  private Session getSession() {
      return sf.getCurrentSession();
  }
```

```
    // unmodified methods omitted
}
```

> **注意：**
> 因为SessionFactory也可以创建EntityManager，所以也可以使用 7.3 小节的 CustomerRepository！

7.5 Spring Data MongoDB

7.5.1 问题

你希望在 Spring Boot 应用程序中使用 MongoDB。

7.5.2 解决方案

将 Mongo Driver 添加为依赖项，并使用 spring.data.mongodb 属性来让 Spring Boot 设置 MongoTemplate 连接到正确的 MongoDB。

7.5.3 工作原理

Spring Boot 自动检测是否存在与 Spring Data MongoDB 类绑定的 MongoDB 驱动程序。如果检查到，那么 Spring Boot 将自动设置 MongoDB 和 MongoTemplate(以及其他相关配置)。

1. 使用 MongoTemplate

现在连接已经建立，你可以使用它了。理想情况下，通过 MongoTemplate 存储和检索文档将更加容易。首先，你需要一个待存储的文档，让我们创建一个希望持久保存的 Customer 类。

```
package com.apress.springboot2recipes.mongo;

public class Customer {
  private String id;

  private final String name;
  private final String email;

  Customer() {
     this(null,null);
  }
```

```
    Customer(String name, String email) {
      this.name = name;
      this.email = email;
    }

    public String getId() {
        return id;
    }

    public String getName() {
        return name;
    }

    public String getEmail() {
        return email;
    }
    // equals(), hashCode() and toString() omitted
}
```

id 字段将自动映射到 MongoDB 的_id 文档标识符。如果想使用其他字段，可以使用来自 Spring Data 的@Id 注解来指定应该使用的字段。

让我们创建 CustomerRepository 接口，以便保存和检索 Customer 实例。

```
package com.apress.springboot2recipes.mongo;

import java.util.List;

public interface CustomerRepository {
  List<Customer> findAll();
  Customer findById(long id);
  Customer save(Customer customer);
}
```

使用 MongoTemplate 实现 MongoDB CustomerRepository 类，如下所示。

```
package com.apress.springboot2recipes.mongo;

import org.springframework.data.mongodb.core.MongoTemplate;
import org.springframework.stereotype.Repository;
```

```
import java.util.List;

@Repository
class MongoCustomerRepository implements CustomerRepository {

  private final MongoTemplate mongoTemplate;

  MongoCustomerRepository(MongoTemplate mongoTemplate) {
      this.mongoTemplate = mongoTemplate;
  }

  @Override
  public List<Customer> findAll() {
      return mongoTemplate.findAll(Customer.class);
  }

  @Override
  public Customer findById(long id) {
      return mongoTemplate.findById(id, Customer.class);
  }

  @Override
  public Customer save(Customer customer) {
    mongoTemplate.save(customer);
    return customer;
  }
}
```

MongoCustomerRepository 使用预先配置的 MongoTemplate 在 MongoDB 中存储和检索客户数据。

要使用上述代码的所有功能，首先 MongoDB 中需要一些数据，可以编写如下 ApplicationRunner 来插入一些数据。

```
@Component
@Order(1)
class DataInitializer implements ApplicationRunner {

  private final CustomerRepository customers;
```

```java
  DataInitializer(CustomerRepository customers) {
    this.customers = customers;
  }

  @Override
  public void run(ApplicationArguments args) throws Exception {

    List.of(
        new Customer("Marten Deinum","marten.deinum@conspect.nl"),
        new Customer("Josh Long", "jlong@pivotal.io"),
        new Customer("John Doe", "john.doe@island.io"),
        new Customer("Jane Doe", "jane.doe@island.io"))
        .forEach(customers::save);
  }
}
```

DataInitializer 类使用 CustomerRepository 实例将一些 Customer 实例保存到 MongoDB 中。请注意上述代码中的@Order 注解，可以通过该注解显式地指定 bean 的执行顺序，这里使用@Order(1)，表示强制首先执行这个 bean。

接下来创建一个 ApplicationRunner 来检索来自 MongoDB 的所有客户数据。

```java
@Component
class CustomerLister implements ApplicationRunner {

  private final Logger logger = LoggerFactory.getLogger(getClass());
  private final CustomerRepository customers;

  CustomerLister(CustomerRepository customers) {
    this.customers = customers;
  }

  @Override
  public void run(ApplicationArguments args) {

    customers.findAll().forEach(customer->logger.info("{}",customer));
  }
}
```

第 7 章 ■ 数 据 访 问

CustomerLister 将使用 CustomerRepository 从数据库加载所有客户数据,并在日志中打印一行信息。

最后,编写应用程序类来引导所有这些类(包括前面编写的两个 ApplicationRunner 类)。

```
package com.apress.springboot2recipes.mongo;

import org.slf4j.Logger;
import org.slf4j.LoggerFactory;
import org.springframework.boot.ApplicationArguments;
import org.springframework.boot.ApplicationRunner;
import org.springframework.boot.SpringApplication;
import org.springframework.boot.autoconfigure.SpringBootApplication;
import org.springframework.core.annotation.Order;
import org.springframework.stereotype.Component;

import java.util.Arrays;

@SpringBootApplication
public class MongoApplication {

  public static void main(String[] args) {
      SpringApplication.run(MongoApplication.class, args);
  }
}

@Component
@Order(1)
class DataInitializer implements ApplicationRunner {
  private final CustomerRepository customers;

  DataInitializer(CustomerRepository customers) {
      this.customers = customers;
  }

  @Override
  public void run(ApplicationArguments args) throws Exception {

    List.of(
```

```
            new Customer("Marten Deinum", "marten.deinum@conspect.nl"),
            new Customer("Josh Long", "jlong@pivotal.io"),
            new Customer("John Doe", "john.doe@island.io"),
            new Customer("Jane Doe", "jane.doe@island.io"))
                .forEach(customers::save);
    }
}

@Component
class CustomerLister implements ApplicationRunner {

    private final Logger logger = LoggerFactory.getLogger(getClass());
    private final CustomerRepository customers;

    CustomerLister(CustomerRepository customers) {
        this.customers = customers;
    }

    @Override
    public void run(ApplicationArguments args) {

        customers.findAll().forEach(customer->logger.info("{}",customer));
    }
}
```

在运行应用程序时，它将自动连接到 MongoDB 实例，向其中插入数据，最后检索数据并将其打印到日志中。

你将需要一个 MongoDB 实例来存储和检索文档。可以使用嵌入式 MongoDB(非常适合测试)，也可以连接到实际的 MongoDB 实例。

2. 使用嵌入式 MongoDB

嵌入式 MongoDB[6] 在 Spring Boot 中是一个易于使用的数据库，当 Spring Boot 启动时，它将自动启动 MongoDB。Spring Boot 甚至通过 spring.mongodb.embedded 名称空间中的属性(如表 7-5 所示)对这个嵌入式的 MongoDB 提供了广泛的支持。使用该数据库唯一需要做的是添加必要的依赖性。

---

6 https://flapdoodle-oss.github.io/de.flapdoodle.embed.mongo/.

```xml
<dependency>
    <groupId>de.flapdoodle.embed</groupId>
    <artifactId>de.flapdoodle.embed.mongo</artifactId>
</dependency>
```

表 7-5 嵌入式 MongoDB 的配置属性

属性	描述
spring.mongodb.embedded.version	所使用的 MongoDB 的版本号。默认为 3.2.2
spring.mongodb.embedded.features	启用的特性清单,默认值仅包含 sync-delay
spring.mongodb.embedded.storage.database-dir	保存数据的目录
spring.mongodb.embedded.storage.oplog-size	日志的大小,以 megabyte 为单位
spring.mongodb.embedded.storage.repl-set-name-dir	数据副本集的名称,默认为 null

现在,当运行 MongoApplication 时,它也会自动启动 MongoDB,并在应用程序停止时将其关闭。

3. 连接外部 MongoDB

■ 注意:
bin目录包含mongo.sh脚本,该脚本将使用Docker启动MongoDB实例。

在启动没有使用嵌入式 MongoDB 的 MongoApplication 时,默认情况下,它将尝试连接到本地主机和端口 27017 上的 MongoDB 服务器。如果你不想连接到该位置,请使用 spring.data.mongodb 属性(参见表 7-6)配置正确的位置。

表 7-6 MongoDB 配置属性

属性	说明
spring.data.mongodb.uri	Mongo 数据库的 URI,包括凭据、设置等。默认值为 mongodb://localhost/test
spring.data.mongodb.username	登录 Mongo 服务器的用户。默认为空
spring.data.mongodb.password	登录 Mongo 服务器的密码。默认为空
spring.data.mongodb.host	MongoDB 服务器的主机名(或 IP 地址)。默认为 localhost
spring.data.mongodb.port	MongoDB 服务器的端口号,默认为 27017
spring.data.mongodb.database	要使用的 MongoDB 数据库/集合的名称
spring.data.mongodb.fieldnaming-strategy	用于将对象字段映射到文档字段的 FieldNamingStrategy 的 FQN。默认为 PropertyNameFieldNamingStrategy

(续表)

属性	说明
spring.data.mongodb.authentication-database	用于身份验证的 MongoDB 的名称。默认为 spring.data.mongodb.database 属性的值
spring.data.mongodb.grid-fs-database	要连接的 MongoDB 的名称。默认为 spring.data.mongodb.database 属性的值

■ 提示：

只有 MongoDB 2.0 支持 spring.data.mongodb.username、spring.data.mongodb.password、spring.data.mongodb.host、spring.data.mongodb.port 和 spring.data.mongodb.database 属性。对于 MongoDB 3.0 及以上版本，必须使用 spring.data.mongodb.uri 属性！

尽管可以使用 Spring Boot 来配置 MongoClient，但如果需要更多的控制，可以将自己的 MongoDbFactory 或 MongoClient 声明为 bean，而 Spring Boot 也不会自动配置作为 bean 的 MongoClient。Spring Boot 仍将检测 Spring Data MongoDB 类，并启用存储库支持。

4. 使用 Spring Data MongoDB 存储库

与编写自己的存储库实现不同，还可以使用 Spring Data MongoDB 实现对数据库的各种操作(就像使用 Spring Data JPA 一样)。为此，你需要扩展 Spring Data 的任意一个 Repository 接口。最简单的方法是扩展 CrudRepository 或 MongoRepository。

```
public interface CustomerRepository extends MongoRepository<Customer, String>
{ }
```

这就是获得一个功能完备的存储库所需要的全部内容，你可以删除前面的 MongoCustomerRepository 的实现。上述代码将为你提供 save、findAll、findById 和许多其他方法。

应用程序仍将运行，插入一些客户数据，并在日志中打印出这些客户数据。

5. 反应式 MongoDB 存储库

除了常规的阻塞操作以外，MongoDB 还支持反应式的使用方式。为此，使用 spring-boot-starter-data-mongodb-reactive 依赖项来代替 spring-boot-starter-data-mongodb 依赖项。这个依赖项将引入 MongoDB 所需的反应式库和反应式驱动。

```xml
<dependency>
  <groupId>org.springframework.boot</groupId>
  <artifactId>spring-boot-starter-data-mongodb-reactive
```

```
</artifactId>
</dependency>
```

在添加正确的依赖关系后，CustomerRepository 将变成反应式的。只需要根据需要扩展 ReactiveRepository、ReactiveSortingRepository 或 ReactiveMongoRepository 即可完成修改。

```
public interface CustomerRepository
    extends ReactiveMongoRepository<Customer, String> { }
```

如上所示，CustomerRepository 类扩展了 ReactiveMongoRepository 类，因此所有方法要么返回 Flux 类型的值(包含零个或多个元素)，要么返回 Mono 类型的值(包含零个或一个元素)。

■ 提示：
默认的实现使用 Project Reactor[7] 作为反应式框架，但是，你也可以使用 RxJava[8]。如果使用的是 RxJava，返回值的类型将变成 Observable 或 Single，而不再是 Flux 和 Mono。为此，需要扩展 RxJava2CrudRepository 或 RxJava2SortingRepository。

DataInitializer 类也需要修改为反应式的。

```
@Component
@Order(1)
class DataInitializer implements ApplicationRunner {

  private final CustomerRepository customers;

  DataInitializer(CustomerRepository customers) {
      this.customers = customers;
  }

  @Override
  public void run(ApplicationArguments args) throws Exception {

      customers.deleteAll()
        .thenMany(
          Flux.just(
            new Customer("Marten Deinum","marten.deinum@conspect.nl"),
```

---

7 https://projectreactor.io.
8 https://github.com/ReactiveX/RxJava.

```
        new Customer("Josh Long", "jlong@pivotal.io"),
        new Customer("John Doe", "john.doe@island.io"),
        new Customer("Jane Doe", "jane.doe@island.io"))
    ).flatMap(customers::save).subscribe(System.out::println);
  }
}
```

首先，代码删除表中的所有内容，然后创建多个新的 Customer 对象并将它们添加到存储库中。最后我们需要订阅这个流，否则什么都不会发生。除了使用 subscribe 方法之外，也可以使用 block 方法，但应该避免在反应式应用程序中使用阻塞式的方法。

最后，还需要将 CustomerListener 修改为反应式的。

```
@Component
class CustomerListener implements ApplicationRunner {

  private final Logger logger = LoggerFactory.getLogger(getClass());
  private final CustomerRepository customers;

  CustomerListener(CustomerRepository customers) {
      this.customers = customers;
  }

  @Override
  public void run(ApplicationArguments args) {

      customers.findAll().subscribe(customer -> logger.info("{}",
      customer));
  }
}
```

上述代码中，findAll 方法将从存储库中获取所有的客户对象，并在每一行打印一个客户的数据。

在启动应用程序时，可能在日志文件中看不到任何内容。由于应用程序的反应式特性，它完成得很快，而 CustomerListener 还没来得及注册自己并开始监听。为了防止关闭应用程序，可以添加代码 System.in.read()，这将使应用程序一直运行，直到按下 Enter 键。通常，运行应用程序时不需要这样做，因为应用程序作为服务/Web 应用公开后将自动运行。

```
public static void main(String[] args) throws IOException {
```

```
    SpringApplication.run(ReactiveMongoApplication.class, args);
    System.in.read();
}
```

现在，当运行应用程序时，应用程序将从 MongoDB 中检索客户数据并打印在日志中。这看起来可能差异不是很大，但是检索的方式是完全不同的。只要对 DataInitializer 稍做修改就可以清楚地说明这一点，对 Customer 的操作将延迟 250 毫秒。

```
@Component
@Order(1)
class DataInitializer implements ApplicationRunner {

    private final CustomerRepository customers;

    DataInitializer(CustomerRepository customers) {
        this.customers = customers;
    }

    @Override
    public void run(ApplicationArguments args) throws Exception {

        customers.deleteAll().thenMany(
            Flux.just(
              new Customer("Marten Deinum","marten.deinum@conspect.nl"),
              new Customer("Josh Long", "jlong@pivotal.io"),
              new Customer("John Doe", "john.doe@island.io"),
              new Customer("Jane Doe", "jane.doe@island.io"))
        ).delayElements(Duration.ofMillis(250))
            .flatMap(customers::save)
            .subscribe(System.out::println);
    }
}
```

现在，插入每个 Customer 对象的操作将延迟 250 毫秒。在运行应用程序时，你将看到 CustomerListener 也需要一些时间来显示每个客户的信息。

6. 测试 Mongo 存储库

在测试 MongoDB 代码时，需要使用一个正在运行的 Mongo 实例，对于测试，通常使用嵌入式 MongoDB。使用@DataMongoTest 注解，在 Spring Boot 中可以很容易地为 MongoDB 编写测试。Spring Boot 将创建一个只包含 MongoDB 相关 bean 的最小化应用程序,而且,如果检测到 MongoDB 是嵌入式的,将自动启动该存储库。

让我们使用一个嵌入式 MongoDB 为 CustomerRepository 编写一个测试。首先,添加嵌入式 MongoDB 的测试依赖项。

```xml
<dependency>
  <groupId>de.flapdoodle.embed</groupId>
  <artifactId>de.flapdoodle.embed.mongo</artifactId>
  <scope>test</scope>
</dependency>
```

接下来创建 CustomerRepositoryTest 类。

```java
@RunWith(SpringRunner.class)
@DataMongoTest
public class CustomerRepositoryTest {

    @Autowired
    private CustomerRepository repository;

    @After
    public void cleanUp() {
        repository.deleteAll();
    }
}
```

@RunWith(SpringRunner.class)通过一个特殊的 JUnit 运行器执行测试,以启动 Spring Test Context 框架。@DataMongoTest 注解将使用嵌入式的 MongoDB(如果在类路径上)替换预配置的 MongoDB。它还将启动 MongoDB 组件,并在检测到 Spring Data Mongo 时启动该存储库。

在每次测试完成之后,我们都希望确保嵌入式 MongoDB 中不再包含任何数据。这可以通过添加带@After 注解的方法来实现,该方法在存储库上调用 deleteAll 方法。这个方法将在每次测试方法执行完成之后被调用。

现在编写一个测试来检查记录的插入是否正确。

```java
@Test
public void insertNewCustomer() {
  assertThat(repository.findAll()).isEmpty();

  Customer customer = repository.save(new Customer(-1, "T. Testing",
  "t.testing@test123.tst"));
```

```java
    assertThat(customer.getId()).isGreaterThan(-1L);
    assertThat(customer.getName()).isEqualTo("T. Testing");
    assertThat(customer.getEmail()).isEqualTo("t.testing@test123.tst");

    assertThat(repository.findById(customer.getId())).isEqualTo
    (customer);
}
```

该测试首先断言数据库是空的——不一定需要这样做，但可以检测其他测试是否污染了数据库。接下来，通过调用 CustomerRepository 类的 save 方法将 Customer 对象插入到数据库中。对插入了数据的 Customer 表格，使用断言验证其 id 字段不为空，以及 name 和 email 属性的值已保存在表格中。最后，再次从 Customer 表格检索数据并与插入前的数据进行比较。

另一个可以测试的是 findAll 方法。当插入两条记录时，调用 findAll 将检索这两条记录。

```java
@Test
public void findAllCustomers() {
    assertThat(repository.findAll()).isEmpty();

    repository.save(new Customer(-1,"T. Testing1",
      "t.testing@test123.tst"));
    repository.save(new Customer(-1, "T. Testing2",
      "t.testing@test123.tst"));

    assertThat(repository.findAll()).hasSize(2);
}
```

当然可以使用更多断言进行其他测试，但数据保存的问题已经在其他测试方法中进行了验证。

第 8 章

Java 企业服务

在本章中，你将了解 Spring 为最常见的 Java 企业服务提供的支持：Java 管理扩展(Java Management eXtensions，JMX)、使用 JavaMail 发送电子邮件、后台处理和任务调度。

JMX 是 JavaSE 的一部分，是一种用于管理和监视系统资源(如设备、应用程序、对象和服务驱动的网络)的技术。这些资源都是系统中的托管 bean(MBean)。Spring 通过将所有 Spring Bean 导出为 MBean 模型来支持 JMX，而不需要针对 JMX API 进行编程。此外，Spring 还可以轻松地访问远程 MBean。

JavaMail 是用于在 Java 中发送电子邮件的标准 API 和实现。Spring 还提供了一个抽象层，以独立于实现的方式发送电子邮件。

8.1 Spring 异步处理机制

8.1.1 问题

你希望使用一个长时间处于运行状态的方法异步调用其他方法。

8.1.2 解决方案

Spring 支持配置 TaskExecutor，并能够异步执行使用@Async 注解的方法。这可以以透明的方式完成，而不需要为异步执行进行常规设置。然而，Spring Boot 不会自动检测应用程序需要执行异步方法，必须使用@EnableAsync 注解启用此特性。

8.1.3 工作原理

让我们编写一个组件，该组件以异步方式将某些内容打印到控制台。

```
package com.apress.springbootrecipes.scheduling;

import org.slf4j.Logger;
```

```java
import org.slf4j.LoggerFactory;
import org.springframework.scheduling.annotation.Async;
import org.springframework.scheduling.annotation.Scheduled;
import org.springframework.stereotype.Component;

@Component
public class HelloWorld {

    private static final Logger logger = LoggerFactory.
    getLogger(HelloWorld.class);

    @Async
    public void printMessage() throws InterruptedException {
        Thread.sleep(500);
        logger.info("Hello World, from Spring Boot 2!");
    }
}
```

HelloWorld 类将等待 500 毫秒，然后将某些内容打印到日志中。注意方法上的 @Async 注解，这表示该方法可以异步执行。但是，必须为 Spring Boot 应用程序显式地启用此特性。

要启用异步处理，需要添加@EnableAsync 配置注解。最简单的解决方案是将这个注解添加到应用程序类上。

```java
package com.apress.springbootrecipes.scheduling;

import org.springframework.boot.ApplicationRunner;
import org.springframework.boot.SpringApplication;
import org.springframework.boot.autoconfigure.SpringBootApplication;
import org.springframework.context.annotation.Bean;
import org.springframework.scheduling.annotation.EnableAsync;
import org.springframework.scheduling.annotation.EnableScheduling;

import java.io.IOException;

@SpringBootApplication
@EnableAsync
public class ThreadingApplication {
```

```
    public static void main(String[] args) throws IOException {
        SpringApplication.run(ThreadingApplication.class, args);

        System.out.println("Press [ENTER] to quit:");
        System.in.read();
    }

    @Bean
    public ApplicationRunner startupRunner(HelloWorld hello) {
        return (args) -> {hello.printMessage();};
    }
}
```

默认情况下，Spring Boot 将创建一个名为 applicationTaskExecutor 的 TaskExecutor 实例。在添加@EnableAsync 注解后，Spring Boot 将自动检测此实例并将其用于方法的异步执行，同时启动检测使用@Async 注解的方法。

代码中的 System.in.read 将防止应用程序关闭，这样后台任务就可以完成处理。当按回车键时，程序将退出。一般来说，当开发一个 Web 应用时，你不需要这样的处理。

1. 配置 TaskExecutor

Spring Boot 将自动配置一个 ThreadPoolTaskExecutor 对象。该对象可以通过 spring.task.execution 名称空间中的属性进行配置，如表 8-1 所示。

表 8-1　Spring Boot Task Executor 属性

属性	说明
spring.task.execution.pool.core-size	核心线程数，默认为 8
spring.task.execution.pool.max-size	最大线程数，默认为 Integer.MAX_VALUE
spring.task.execution.pool.queue-capacity	队列容量。默认无限大，如果设置了该属性，将忽略 max-size 属性的值
spring.task.execution.pool.keep-alive	保留空闲线程的时间上限，超过时限的空闲线程将被终止
spring.task.execution.thread-name-prefix	新建线程的名称前缀。默认是 task-
spring.task.execution.pool.allow-core-thread-timeout	是否允许核心线程超时，默认为 true。这使得线程池能够动态增长和收缩

向 application.properties 配置文件中添加以下内容将覆盖一些默认设置。

```
spring.task.execution.pool.core-size=4
```

```
spring.task.execution.pool.max-size=16
spring.task.execution.pool.queue-capacity=125
spring.task.execution.thread-name-prefix=sbr-exec-
```

现在，在重新运行应用程序时，线程的名称将以 sbr-exec 作为前缀。它将启动 4 个线程，并将增加到最多 16 个。要使线程池能自动调整大小，需要为队列的容量提供一个固定的数字。太大的数字(或无边界)将不会使线程池自动增加线程的数量。

2. 使用 TaskExecutorBuilder 创建 TaskExecutor

如果需要构造 TaskExecutor 对象，可以使用 Spring Boot 提供的 TaskExecutorBuilder 类。这个生成器类使得构造 ThreadPoolTaskExecutor 变得更容易。它允许我们设置与表 8-1 中相同的属性。

```
@Bean
public TaskExecutor customTaskExecutor(TaskExecutorBuilder builder) {
    return builder.corePoolSize(4)
                  .maxPoolSize(16)
                  .queueCapacity(125)
                  .threadNamePrefix("sbr-exec-").build();
}
```

如果由于某种原因，应用程序中存在多个 TaskExecutor 实例，则需要将其中一个实例标记为@Primary 以用作默认的 TaskExecutor 实例，或者使用 AsyncConfigurer 接口并实现 taskExecutor 方法以返回默认的 TaskExecutor 实例。

```
@SpringBootApplication
@EnableAsync
public class ThreadingApplication implements AsyncConfigurer {

  @Bean
  public ThreadPoolTaskExecutor taskExecutor() { ... }

  @Override
  public Executor getAsyncExecutor() {
    return taskExecutor();
  }
}
```

8.2 Spring 任务调度

8.2.1 问题

你希望使用 cron 表达式、间隔或速率以一致的方式安排方法调用。

8.2.2 解决方案

Spring 支持配置多个 TaskExecutor 和 TaskScheduler。这种功能，加上使用 @Scheduled 注解来调度方法执行的能力，使得在 Spring 中实现任务调度所需的开发工作非常少：只需要一个方法、一个注解，然后打开注解扫描器来读取注解即可。Spring Boot 不会自动检测到应用程序需要进行任务调度，必须使用@EnableScheduling 注解来启用任务调度。

8.2.3 工作原理

让我们编写一个组件，每隔四秒钟在日志中打印出一条消息。创建一个 Java 类，在该类的一个方法上使用@Scheduled 注解来指示该方法需要被触发。使用 fixedRate=4000 作为参数，它将每四秒运行一次。如果希望使用 cron 表达式，可以修改为在@Scheduled 注解上设置 cron 属性。

```java
package com.apress.springbootrecipes.scheduling;

import org.slf4j.Logger;
import org.slf4j.LoggerFactory;
import org.springframework.scheduling.annotation.Scheduled;
import org.springframework.stereotype.Component;

@Component
public class HelloWorld {

  private static final Logger logger =
    LoggerFactory.getLogger(HelloWorld.class);

  @Scheduled(fixedRate = 4000)
  public void printMessage() {
      logger.info("Hello World, from Spring Boot 2!");
  }
}
```

@Component 将确保 Spring Boot 会检测到这个类。

接下来要做的是为应用程序启用任务调度。最简单的解决方案是使用 @EnableScheduling 来注解应用程序类。当然，也可以将它放在其他@Configuration 注解的类上。

```
package com.apress.springbootrecipes.scheduling;

import org.springframework.boot.SpringApplication;
import org.springframework.boot.autoconfigure.SpringBootApplication;
import org.springframework.scheduling.annotation.EnableScheduling;

@SpringBootApplication
@EnableScheduling
public class SchedulingApplication {

  public static void main(String[] args) {
      SpringApplication.run(SchedulingApplication.class, args);
  }
}
```

@EnableScheduling 将使 Spring Boot 检测到@Scheduled 注解的方法，并将注册一个 TaskScheduler 实例用于调度任务。当在应用程序上下文中检测到单个 TaskScheduler 实例时，应用程序将使用该任务调度器而不是创建一个新的实例。

运行 SchedulingApplication 时，应用程序大约每四秒钟就会在日志中打印一条记录。

除了使用@Scheduled 注解以外，还可以使用 Java 调度一个方法。在无法将 @Scheduled 注解添加到需要定期执行的方法上，或者只是想限制注解的数量时，就需要使用这种方法。为此，可以使用 SchedulingConfigurer 接口，它有一个回调方法来配置其他任务。

```
@SpringBootApplication
@EnableScheduling
public class SchedulingApplication implements SchedulingConfigurer {

  @Autowired
  private HelloWorld helloWorld;

  public static void main(String[] args) {
      SpringApplication.run(SchedulingApplication.class, args);
  }
```

```
    @Override
    public void configureTasks(ScheduledTaskRegistrar taskRegistrar) {
        taskRegistrar.addFixedRateTask(
                () -> helloWorld.printMessage()
                , 4000);
    }
}
```

8.3 发送 E-mail

当在类路径上检测到邮件相关的属性和 Java 的邮件库时，Spring Boot 将自动配置发送邮件的功能。在本节中，我们将了解如何设置这些属性以及如何使用 Spring Boot 发送电子邮件。

8.3.1 问题

你希望从 Spring Boot 应用程序发送电子邮件。

8.3.2 解决方案

Spring 提供了一个抽象的、与实现无关的用于发送电子邮件的 API，使得发送电子邮件更加容易。Spring 电子邮件特性的核心接口是 MailSender。JavaMailSender 接口是 MailSender 的一个子接口，它包括专用的 JavaMail 特性，如多用途 Internet 邮件扩展(Multipurpose Internet Mail Extensions，简称 MIME 或 MIME 消息)支持。要发送带有 HTML 内容、内嵌图像或附件的电子邮件，必须将其作为 MIME 消息发送。当在类路径中找到 javax.mail 类以及设置了适当的 spring.mail 属性时，Spring Boot 将自动配置 JavaMailSender。

8.3.3 工作原理

首先要做的是将 spring-boot-starter-mail 依赖项添加到依赖项清单中。这将在类路径上添加必要的 javax.mail 库以及 spring-context 依赖项。

```
<dependency>
  <groupId>org.springframework.boot</groupId>
  <artifactId>spring-boot-starter-mail</artifactId>
</dependency>
```

1. 配置 JavaMailSender

为了能够发送邮件，需要配置适当的 spring.mail 属性，如表 8-2 所示。至少需要配置 spring.mail.host 属性，其他属性是可选的。

表 8-2 Spring Boot 邮件属性

属性	说明
spring.mail.host	SMTP 服务器主机
spring.mail.port	SMTP 服务器端口(默认端口为 25)
spring.mail.username	连接 SMTP 服务器的用户名
spring.mail.password	连接 SMTP 服务器的用户密码
spring.mail.protocol	SMTP 服务器使用的协议(默认协议是 smtp)
spring.mail.test-connection	是否在启动时检测 SMTP 服务器的可用性(默认为 false)
spring.mail.default-encoding	MIME 消息的编码(默认编码为 UTF-8)
spring.mail.properties.*	JavaMail Session 上可配置的其他属性
spring.mail.jndi-name	JavaMail Session 的 JNDI 名称；在将 JavaMail Session 部署到 JEE 服务器时，可以在 JNDI 中预先配置该 JavaMail Session 的 JNDI 名称

接下来，至少需要设置 spring.mail.host 属性才能发送邮件。

```
spring.mail.host=localhost
spring.mail.port=3025
```

■ 注意：
本小节的代码使用 GreenMail 作为 SMTP 服务器，可以使用 bin 目录中的 smtp.sh 脚本运行配置好的实例。默认情况下，它将在端口 3025 上开放 SMTP 服务器。

2. 发送纯文本 E-mail

在添加了依赖项和设置相关的 spring.mail 属性之后，Spring Boot 将向 ApplicationContext 添加一个预配置的 JavaMailSenderImpl 作为 bean。这个 bean 可以通过使用@Autowired 字段，或通过构造函数，或如以下代码所示，作为@Bean 注解的方法中的参数，自动填充到组件中。

```
package com.apress.springbootrecipes.mailsender;

import org.springframework.boot.ApplicationRunner;
import org.springframework.boot.SpringApplication;
import org.springframework.boot.autoconfigure.SpringBootApplication;
import org.springframework.context.annotation.Bean;
import org.springframework.mail.javamail.JavaMailSender;
import org.springframework.mail.javamail.MimeMessageHelper;
```

```
import javax.mail.Message;

@SpringBootApplication
public class MailSenderApplication {

    public static void main(String[] args) {
        SpringApplication.run(MailSenderApplication.class, args);
    }

    @Bean
    public ApplicationRunner startupMailSender(JavaMailSender mailSender) {
        return (args) -> {
            mailSender.send((msg) -> {
                var helper = new MimeMessageHelper(msg);
                helper.setTo("recipient@some.where");
                helper.setFrom("spring-boot-2-recipes@apress.com");
                helper.setSubject("Status message");
                helper.setText("All is well.");
            });
        };
    }
}
```

在应用程序启动后，MailSenderApplication 将发送一封电子邮件。startupMailSender 方法是一个 ApplicationRunner(参见第 2 章)，它接收预先配置的 JavaMailSender 以发送邮件消息。

3. 使用 Thymeleaf 作为 E-mail 模板

Spring Boot 对使用 Thymeleaf[1] 作为模板化解决方案提供了一些很好的支持，然而，默认设置主要是将 Thymeleaf 用于网页。但是，完全可以将 Thymeleaf 作为电子邮件的模板。

首先，添加 spring-boot-starter-thymeleaf 依赖项。这将引入所有需要的 Thymeleaf 依赖项，并自动配置 Thymeleaf 模板引擎，我们需要它来生成 HTML 内容。

```
<dependency>
```

---
[1] https://www.thymeleaf.org.

```xml
    <groupId>org.springframework.boot</groupId>
    <artifactId>spring-boot-starter-thymeleaf</artifactId>
</dependency>
```

默认情况下，Spring 配置的 Thymeleaf TemplateEngine 将从 src/main/resources 下的 templates 目录中解析 HTML 模板。在该目录下添加名为 email.html 的文件，并根据它生成一个美观的电子邮件消息。

```html
<!DOCTYPE html>
<html xmlns:th="http://www.thymeleaf.org">
<head>
    <meta http-equiv="Content-Type" content="text/html; charset=UTF-8" />
</head>
<body>
<p><strong th:text="${msg}">Some email content will be here.</strong></p>
<p>
Kind Regards,
    Your Application
</p>
</body>
</html>
```

th:text 是一个 Thymeleaf 标签，它将用该属性的值替换内容。当然，我们将需要从邮件发送/生成代码中传入该属性的值。

```java
package com.apress.springbootrecipes.mailsender;

import org.springframework.boot.ApplicationRunner;
import org.springframework.boot.SpringApplication;
import org.springframework.boot.autoconfigure.SpringBootApplication;
import org.springframework.context.annotation.Bean;
import org.springframework.context.i18n.LocaleContextHolder;
import org.springframework.mail.javamail.JavaMailSender;
import org.springframework.mail.javamail.MimeMessageHelper;
import org.thymeleaf.context.Context;
import org.thymeleaf.spring5.SpringTemplateEngine;

import javax.mail.Message;
```

```
import java.util.Collections;

@SpringBootApplication
public class MailSenderApplication {

  public static void main(String[] args) {
     SpringApplication.run(MailSenderApplication.class, args);
  }

  @Bean
  public ApplicationRunner startupMailSender(
      JavaMailSender mailSender,
      SpringTemplateEngine templateEngine) {
    return (args) -> {
       mailSender.send((msg) -> {
          var helper = new MimeMessageHelper(msg);
          helper.setTo("recipient@some.where");
          helper.setFrom("spring-boot-2-recipes@apress.com");
          helper.setSubject("Status message");

          var context = new Context(
              LocaleContextHolder.getLocale(),
              Collections.singletonMap("msg","All is well!"));
          var body=templateEngine.process("email.html", context);
          helper.setText(body, true);
       });
    };
  }
}
```

上述代码仍然与前面的代码非常相似，但是现在使用了 Spring TemplateEnginge 来为我们的电子邮件生成 HTML 内容。代码中使用 process 方法来选择我们希望呈现的模板，即 email.html，并传入一个 Context 对象。Thymeleaf 使用该 Context 对象来解析属性，在上述示例中解析的是 msg。

8.4 注册 JMX MBean

8.4.1 问题

你希望在 Spring Boot 应用程序中将对象注册为 JMX MBean，以便能够查看正在运行的服务并在运行时操纵它们的状态。这将允许你执行诸如重新运行批处理作业、调用方法和更改配置元数据之类的任务。

8.4.2 解决方案

Spring Boot 默认启用了 Spring JMX 特性，它将检测 @ManagedResource 注解的 bean，并将其注册到 JMX 服务器。

8.4.3 工作原理

首先，检查一下 Spring Boot 默认启用的 JMX 特性。让我们创建一个简单的 Spring Boot 应用程序，它将保持运行状态，并使用 JConsole 检查这个正在运行的应用程序。

```java
package com.apress.springbootrecipes.jmx;

import org.springframework.boot.SpringApplication;
import org.springframework.boot.autoconfigure.SpringBootApplication;

import java.io.IOException;

@SpringBootApplication
public class JmxApplication {

  public static void main(String[] args) throws IOException {
    SpringApplication.run(JmxApplication.class, args);

    System.out.print("Press [ENTER] to quit:");
    System.in.read();
  }
}
```

当应用程序正在运行时，你可以启动 jconsole 工具，系统会打开一个窗口，可以在其中选择要连接的本地进程。选择运行 JmxApplication 的进程，如图 8-1 所示。

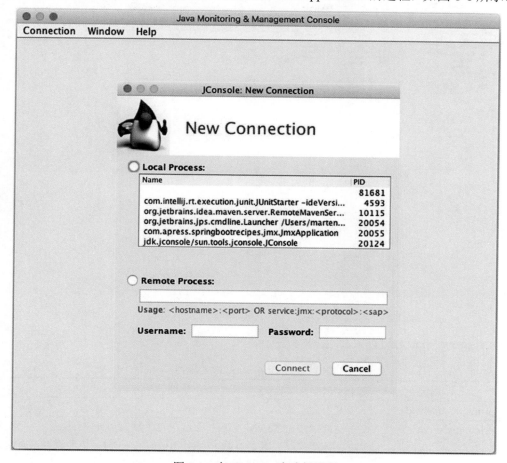

图 8-1　在 JConsole 中选择进程

选择进程后，进入 MBeans 选项卡，在屏幕的左侧展开 org.springframework.boot 菜单以及它下面的所有内容。你可以触发 SpringApplication 上的 shutdown 操作。当触发此操作时，它将关闭应用程序，如图 8-2 所示。

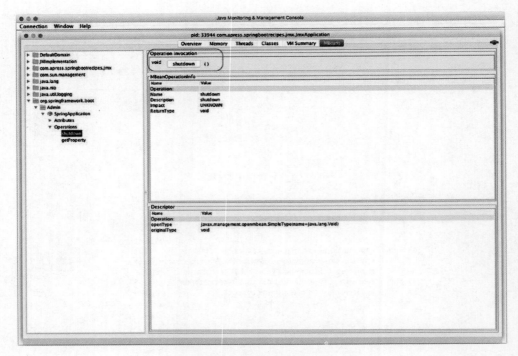

图 8-2　触发 shutdown 操作

Spring Boot 为配置 JMX 提供了三个属性，如表 8-3 所示。

表 8-3　Spring Boot JMX 属性

属性	说明
spring.jmx.enabled	是否启用 JMX，默认为 true
spring.jmx.server	JMX MBeanServer 使用的 bean 名称，默认为 mbeanServer。只有在应用程序上下文中手动注册 MBeanServer 时才需要设置该属性
spring.jmx.default-domain	注册 bean 时使用的 JMX 域名，默认为包名

由于 Spring Boot 默认启用了 JMX 和 Spring 对 JMX 的支持，因此公开 bean 非常简单。对于需要公开的 bean，使用 @ManagedResource 进行注解；对需要公开的操作，使用 @ManagedOperation 进行注解。

```
package com.apress.springbootrecipes.jmx;

import org.slf4j.Logger;
import org.slf4j.LoggerFactory;
import org.springframework.jmx.export.annotation.ManagedOperation;
import org.springframework.jmx.export.annotation.ManagedResource;
```

```
import org.springframework.scheduling.annotation.Async;
import org.springframework.stereotype.Component;

@Component
@ManagedResource
public class HelloWorld {

  private static final Logger logger =
      LoggerFactory.getLogger(HelloWorld.class);

  @ManagedOperation
  public void printMessage() {
      logger.info("Hello World, from Spring Boot 2!");
  }
}
```

当重新启动应用程序,JConsole 重新连接到运行 JmxApplication 的进程时,可以在左侧菜单中看到 com.apress.springbootrecipes.jmx 节点。展开该节点,在 Operations 下可以看到 printMessage 操作,如图 8-3 所示。

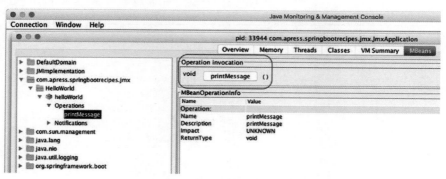

图 8-3　触发 printMessage 操作

在调用 printMessage 方法时,控制台将在日志中打印代码中预置的消息,如图 8-4 所示。

图 8-4　控制台打印的日志

第 9 章

消息传递

9.1 配置 JMS

9.1.1 问题

你希望在 Spring Boot 应用程序中使用 JMS，并且需要连接到 JMS 代理。

9.1.2 解决方案

Spring Boot 支持自动配置 ActiveMQ[1] 和 Artemis[2]，只需要添加这两个 JMS 提供商程序中的一个，并相应地在 spring.activemq 或 spring.artemis 名称空间中设置一些属性即可。

9.1.3 工作原理

通过声明所选 JMS 提供商程序的依赖关系，Spring Boot 将自动为开发环境配置 ConnectionFactory 和查询目的地的策略，即 DestinationResolver。这也可以通过使用 JNDI 来实现。最后一种解决方案是自己完成所有配置，如果希望对 ConnectionFactory 的构造进行更多控制的话。

为 Spring Boot 支持的 JMS 提供商程序添加依赖关系非常容易，只需要引入相应的 spring-boot-starter-*依赖项即可(根据需要，把*替换为 activemq 或 artemis)。对于 JNDI，你需要自己引入 JMS 依赖项。

1. 使用 ActiveMQ

在使用 ActiveMQ 时，首先要做的是添加 spring-boot-starter-activemq 依赖项，这将引入启动 JMS 和 ActiveMQ 所有需要的依赖关系。它将引入 spring-jms 依赖项

---

[1] https://activemq.apache.org.

[2] https://activemq.apache.org/artemis/.

和 ActiveMQ 的客户端库。

```xml
<dependency>
  <groupId>org.springframework.boot</groupId>
  <artifactId>spring-boot-starter-activemq</artifactId>
</dependency>
```

默认情况下，如果未提供显式的代理配置，那么 Spring Boot 将启动嵌入式代理。可以通过使用 spring.activemq 名称空间中的属性修改配置，如表 9-1 所示。

表 9-1 ActiveMQ 配置属性

属性	说明
spring.activemq.broker-url	要连接的代理的 URL，默认情况下，如果是内存内代理，URL 是 vm://localhost?broker.persistent=false，否则是 tcp://localhost:61616
spring.activemq.user	连接到代理的用户名，默认为空
spring.activemq.password	连接到代理的用户密码，默认为空
spring.activemq.in-memory	是否使用嵌入式代理，默认为 true。当显式设置 spring.activemq.broker-url 时忽略该属性的值
spring.activemq.non-blocking-redelivery	在重新发送回滚消息之前停止消息传递。如果启用，则不保留消息的顺序，默认为 false
spring.activemq.close-timeout	认为关闭生效的等待时间，默认为 15 秒
spring.activemq.send-timeout	等待代理响应的时间，默认为 0(表示无时间限制)
spring.activemq.packages.trust-all	当使用 Java 序列化发送 JMS 消息时，是否信任所有包中的类，默认为空(需要显式地设置 spring.activemq.packages.trusted 属性)
spring.activemq.packages.trusted	信任的特定包清单，包名之间用逗号分隔

下面是一个简单的应用程序，它列出了名称中包含"jms"的所有 bean，其中应该包括一个名为 cachingJmsConnectionFactory 的 bean。

```java
@SpringBootApplication
public class JmsActiveMQApplication {

  private static final String MSG = "\tName: %100s, Type: %s\n";

  public static void main(String[] args) {
    var ctx=SpringApplication.run(JmsActiveMQApplication.class,args);

    System.out.println("# Beans: " + ctx.getBeanDefinitionCount());
```

```
var names = ctx.getBeanDefinitionNames();
Stream.of(names)
    .filter(name -> name.toLowerCase().contains("jms"))
    .forEach(name -> {
        Object bean = ctx.getBean(name);
        System.out.printf(MSG, name, bean.getClass()
          .getSimpleName());
    });
}
}
```

在运行上述代码时，它会将名称中包含了"jms"的所有 bean 的名称和类型打印到控制台。输出应与图 9-1 相似。

图 9-1　ActiveMQ bean 清单

2. 使用 Artemis

使用 Artemis 时，首先要做的是添加 spring-boot-starter-artemis 依赖项。这将引入启动 JMS 和 Artemis 需要的所有依赖关系。它将引入 spring-jms 依赖项和 Artemis 的客户端库。见表 9-2。

```
<dependency>
  <groupId>org.springframework.boot</groupId>
  <artifactId>spring-boot-starter-artemis</artifactId>
</dependency>
```

表 9-2　Artemis 配置属性

属性	说明
spring.artemis.host	连接 Artemis 代理的主机名，默认为 localhost
spring.artemis.port	连接 Artemis 代理的端口号，默认为 61616
spring.artemis.user	连接 Artemis 代理的用户名，默认为空

(续表)

属性	说明
spring.artemis.password	连接 Artemis 代理的用户密码，默认为空
spring.artemis.mode	操作模式，可以是 native 或 embedded；默认为空，使得 Spring Boot 自动检测模式设置。当找到嵌入式类时，将以 embedded 模式运行

下面是一个简单的应用程序，它列出了名称中包含"jms"的所有 bean，其中应该包括一个名为 cachingJmsConnectionFactory 的 bean。

```
@SpringBootApplication
public class JmsActiveMQApplication {

  private static final String MSG = "\tName: %100s, Type: %s\n";

  public static void main(String[] args) {
    var ctx = SpringApplication.run(JmsActiveMQApplication.class,
      args);

    System.out.println("# Beans: " + ctx.getBeanDefinitionCount());

    var names = ctx.getBeanDefinitionNames();
    Stream.of(names)
      .filter(name -> name.toLowerCase().contains("jms"))
      .sorted(Comparator.naturalOrder())
      .forEach(name -> {
        Object bean = ctx.getBean(name);
        System.out.printf(MSG, name, bean.getClass().
        getSimpleName());
      });
  }
}
```

运行上述代码，输出应与图 9-2 相似。

第 9 章 ■ 消 息 传 递

图 9-2 Artemis bean 清单

■ **注意：**

在使用Artemis时，你可能会想知道配置中是否也有ActiveMQConnectionFactory。Artemis是基于ActiveMQ开发的，因此它与ActiveMQ共享很多类。

可以在嵌入式模式下使用 Artemis(就像 ActiveMQ 一样)，在此模式下它将启动一个嵌入式代理。可以使用在 spring.artemis.embedded 名称空间中公开的几个属性(如表 9-3 所示)对嵌入式代理进行配置。使用嵌入式模式要求添加 artemis-server 依赖项。

```xml
<dependency>
  <groupId>org.apache.activemq</groupId>
  <artifactId>artemis-server</artifactId>
</dependency>
```

表 9-3 Artemis 嵌入式模式配置属性

属性	说明
spring.artemis.embedded.enabled	是否启用嵌入式模式，默认为 true
spring.artemis.embedded.persistent	消息是否持久化，默认为 false
spring.artemis.embedded.data-directory	用于存储日记的目录，仅当 spring.artemis.embedded.persistent 属性设置为 true 时该属性的设置才有用。默认目录为 Java Tempdir
spring.artemis.embedded.queues	启动时创建的队列清单，多个清单名称之间以逗号分隔
spring.artemis.embedded.topics	启动时创建的主题清单，多个主题名称之间以逗号分隔
spring.artemis.embedded.cluster-password	集群密码，默认情况下系统会自动生成密码

3. JNDI

在将 Spring Boot 应用程序部署到 JEE 容器时，你也很可能希望使用该容器中预先注册的 ConnectionFactory。为此，需要依赖于 spring-jms 库和 javax.jms-api(后者可能会被标记，并由你的 JEE 容器提供)。可以通过 spring-boot-starter-*(其中*为 activemq 或 artemis)显式地排除 ActiveMQ 或 Artemis 依赖项；但是，更便捷、更清晰的做法是仅声明所需的依赖项。如下代码引入 spring-jms 和 javax.jms-api 依赖项。

```xml
<dependency>
  <groupId>org.springframework</groupId>
  <artifactId>spring-jms</artifactId>
</dependency>
<dependency>
  <groupId>javax.jms</groupId>
  <artifactId>javax.jms-api</artifactId>
  <scope>provided</scope>
</dependency>
```

当 JNDI(Java Naming and Directory Interface，Java 命名和目录接口)可用时，Spring Boot 将首先尝试使用两个众所周知的名称——java:/JmsXA 和 java:/XAConnectionFactory 中的一个，或者使用由 spring.jms.jndi-name 属性指定的名称，在 JNDI 注册表中检测 ConnectionFactory。此外，它还将自动创建 JndiDestinationResolver，以便在 JNDI 中检测队列和主题。默认情况下，JNDI 允许回退使用动态创建的目的地，通过 DynamicDestinationResolver 提供地址解析。

```
spring.jms.jndi-name=java:/jms/connectionFactory
```

如上述代码所示设置好 spring.jms.jndi-name 属性，并且编译好 WAR 文件(参见 11.2 小节)，现在可以将应用程序部署到 JEE 容器并重用已有的 ConnectionFactory。

4. 手动配置

配置 JMS 的最后一种方法是进行手动配置。为此，至少需要 spring-jms 和 javax.jms-api 依赖项，可能还需要一些用于你使用的 JMS 代理的客户端库。在以下情况下，可能需要手动配置：

(1) Spring Boot 无法自动配置 ConnectionFactory。
(2) 需要对 ConnectionFactory 进行额外的设置。
(3) 需要多个 ConnectionFactory 实例。

为了配置 ConnectionFactory，可以添加一个用@Bean 注解的构造函数。

```java
@Bean
public ConnectionFactory connectionFactory() {
  var connectionFactory = new ActiveMQConnectionFactory
```

```
  ("vm://localhost?
  broker.persistent=false");
  connectionFactory.setClientID("someId");
  connectionFactory.setCloseTimeout(125);
  return connectionFactory;
}
```

上述构造函数将为 ActiveMQ 创建一个 ConnectionFactory 实例，创建过程中使用了一个嵌入式的、非持久化的代理，并设置了 clientId 和 closeTimeout。Spring Boot 检查到预配置的 ConnectionFactory，没有尝试自行创建 ConnectionFactory 实例。

9.2 使用 JMS 发送消息

9.2.1 问题

你希望使用 JMS 向其他系统发送消息。

9.2.2 解决方案

使用 Spring Boot 提供的 JMSTemplate 发送和(根据需要)转换消息。

9.2.3 工作原理

在使用 Spring Boot 时，如果它检测到 JMS 和单个 ConnectionFactory，它还将自动配置一个可用于发送和转换消息的 JmsTemplate。Spring Boot 在 spring.jms.template 名称空间中公开了可用于配置 JmsTemplate 的属性。

1. 通过 JmsTemplate 发送消息

要通过 JMS 发送消息，可以调用 JmsTemplate 的 send 或 sendAndConvert 方法。让我们编写一个组件，该组件每秒钟将带有当前日期和时间的消息放置在队列中。

```
@Component
class MessageSender {

  private final JmsTemplate jms;

  MessageSender(JmsTemplate jms) {
     this.jms = jms;
  }

  @Scheduled(fixedRate = 1000)
```

```
    public void sendTime() {
        jms.convertAndSend("time-queue", "Current Date & Time is: " +
            LocalDateTime.now());
    }
}
```

JmsTemplate 通过构造函数自动注入；由于@Scheduled 注解设置属性 fixedRate = 1000，我们将在名为 time-queue 的队列上每秒收到一条消息，其中包含当前的日期和时间。要运行此代码，你需要一个带有@EnableScheduling 注解的@SpringBootApplication 类，以便处理@Scheduled 注解的方法。

```
@SpringBootApplication
@EnableScheduling
public class JmsSenderApplication {

    public static void main(String[] args) {
        SpringApplication.run(JmsSenderApplication.class, args);
    }
}
```

现在当运行这个类时，看起来好像没有发生什么事情，但是消息正在不断地填入队列中。我们可以编写一个简单的集成测试来检查这段代码是否正常工作。

```
package com.apress.springbootrecipes.demo;

import org.junit.Test;
import org.junit.runner.RunWith;
import org.springframework.beans.factory.annotation.Autowired;
import org.springframework.boot.test.context.SpringBootTest;
import org.springframework.jms.core.JmsTemplate;
import org.springframework.test.context.junit4.SpringRunner;

import javax.jms.JMSException;
import javax.jms.Message;
import javax.jms.TextMessage;

import static org.assertj.core.api.Assertions.assertThat;

@RunWith(SpringRunner.class)
@SpringBootTest
```

```java
public class JmsSenderApplicationTest {

    @Autowired
    private JmsTemplate jms;

    @Test
    public void shouldSendMessage() throws JMSException {

        Message message = jms.receive("time-queue");

        assertThat(message)
                .isInstanceOf(TextMessage.class);
        assertThat(((TextMessage) message).getText())
                .startsWith("Current Date & Time is: ");
    }
}
```

上述 JUnit 测试将启动应用程序并开始发送消息。在测试中，我们使用 JmsTemplate 的 receive 方法接收消息，并使用断言查看消息是否被发送并包含我们期望它们包含的内容。测试将使用嵌入式 JMS 代理，因为我们没有配置其他代理。在运行测试时，它应该一直处于正常运行状态，因为消息在不断地发送和接收。

■ 提示：

在为消息传递编写测试时，你可能希望设置 JmsTemplate 的 receive-timeout 属性，因为默认设置是无时间限制地等待消息的到来。然而，你可能希望在等待 500 毫秒之后还没有收到消息时认为测试失败。可以通过在 application.properties 文件中添加 spring.jms.template.receive-timeout=500ms 完成设置。

2. 配置 JmsTemplate

Spring Boot 在 spring.jms.template 名称空间中提供了配置 JmsTemplate 的属性，如表 9-4 所示。

表 9-4 JmsTemplate 属性

属性	说明
spring.jms.template.default-destination	未指定特定目标时用于发送和接收操作的默认目标
spring.jms.template.delivery-delay	发送消息的传递延迟
spring.jms.template.delivery-mode	发送模式，可以设置为 persistent 或 non-persistent；当显式设置该属性的值时，需要将 qos-enabled 属性设置为 true

(续表)

属性	说明
spring.jms.template.priority	发送时消息的优先级。默认为空，当显式设置该属性的值时，需要将 qos-enabled 属性设置为 true
spring.jms.template.qos-enabled	是否应启用 QoS(Quality of Service，服务质量)。如果启用，则必须设置 delivery-mode 和 time-to-live 属性。默认值为 false
spring.jms.template.receive-timeout	接收消息的超时时间。默认无限制
spring.jms.template.time-to-live	JMS 消息的生存时间；当显式设置该属性的值时，需要将 qos-enabled 属性设置为 true
spring.jms.pub-sub-domain	默认目标是主题还是队列。默认为 false，表示是队列

除了使用这些属性以外，在仅有一个 JmsTemplate bean 时，还可以通过 DestinationResolver 和 MessageConverter 自动配置 JmsTemplate。如果不存在唯一的 JmsTemplate 实例，将使用 DynamicDestinationResolver 和 SimpleMessageConverter 类中的属性(如表 9-5 所示)的默认值配置 JmsTemplate。

表 9-5　SimpleMessageConverter 和 JMS 消息转换器的类型对应关系

数据类型	JMS 消息类型
java.lang.String	javax.jms.TextMessage
java.util.Map	javax.jms.MapMessage
java.io.Serializable	javax.jms.ObjectMessage
byte[]	javax.jms.BytesMessage

让我们向 orders 队列发送一个 Order 对象。使用 Jackson 而不是 Java 序列化机制来发送 JSON。

```java
public class Order {

    private String id;
    private BigDecimal amount;

    public Order() {
    }

    public Order(String id, BigDecimal amount) {
        this.id=id;
        this.amount = amount;
    }
```

```
// Getters / Setters omitted for brevity

  @Override
  public String toString() {
    return String.format("Order[id='%s',amount=%4.2f]",id,amount);
  }
}
```

上述代码创建了简单的要发送的订单对象。

现在，我们需要一个发送者类，这个类构造 Order 对象，并使用 JmsTemplate 将其放置到队列中。

```
@Component
class OrderSender {

  private final JmsTemplate jms;

  OrderSender(JmsTemplate jms) {
      this.jms = jms;
  }

  @Scheduled(fixedRate = 1000)
  public void sendTime() {
      var id = UUID.randomUUID().toString();
      var amount = ThreadLocalRandom.current().nextDouble(1000.00d);
      var order = new Order(id, BigDecimal.valueOf(amount));
      jms.convertAndSend("orders", order);
  }
}
```

这与前面编写的 MessageSender 没有什么区别，但现在它使用一些随机生成的数据创建了一个 Order 对象，并放入 orders 队列发送出去。当运行上述代码时，它实际上会失败。转换无法进行，因为 Order 没有实现 Serializable 接口，而要将 Order 对象转换为 ObjectMessage 对象必须实现该接口(参见表 9-5)。但是，我们希望使用 JSON 实现转换，为此需要一个不同的 MessageConverter，通过 MappingJackson2MessageConverter 实现具体的转换。这个方法使用 Jackson 将对象编组和解组到 JSON，应该将该方法作为 bean 添加到配置中。

首先，需要将 Jackson 依赖项添加到构建配置中。

```xml
<dependency>
  <groupId>com.fasterxml.jackson.core</groupId>
  <artifactId>jackson-core</artifactId>
</dependency>
```

接下来，可以配置 MappingJackson2MessageConverter 方法。

```java
@SpringBootApplication
@EnableScheduling
public class JmsSenderApplication {

    public static void main(String[] args) {
        SpringApplication.run(JmsSenderApplication.class, args);
    }

    @Bean
    public MappingJackson2MessageConverter messageConverter() {
        var messageConverter = new MappingJackson2MessageConverter();
        messageConverter.setTypeIdPropertyName("content-type");
        messageConverter.setTypeIdMappings(
                Collections.singletonMap("order", Order.class));
        return messageConverter;
    }
}
```

typeIdPropertyName 是一个必需的属性，指示存储消息实际类型的属性的名称。如果没有任何进一步的配置，将使用类的 FQN。使用 typeIdMappings，可以指定要将哪种类型映射到哪个类，或者哪个类映射到哪种类型。如果未指定，则将使用类的 FQN 作为映射中的类型。发送 Order 对象时，content-type 头将包含值 order。

■ 提示：
通常，最好显式地定义类型映射。这样，你就不会在Java的级别上显式地绑定两个或多个应用程序。它们可以使用自己的order类型映射到自己的Order类。

做好上述准备之后，我们就可以编写一个测试来查看订单是否被发送了。

```java
@RunWith(SpringRunner.class)
@SpringBootTest
public class JmsSenderApplicationTest {
```

```java
@Autowire
private JmsTemplate jms;

@Test
public void shouldReceiveOrderPlain() throws Exception {

  Message message = jms.receive("orders");

  assertThat(message)
        .isInstanceOf(BytesMessage.class);

  BytesMessage msg = (BytesMessage) message;
  ObjectMapper mapper = new ObjectMapper();
  byte[] content = new byte[(int) msg.getBodyLength()];
  msg.readBytes(content);
  Order order = mapper.readValue( content, Order.class);
  assertThat(order).hasNoNullFieldsOrProperties();
}

@Test
public void shouldReceiveOrderWithConversion() throws Exception {

  Order order = (Order) jms.receiveAndConvert("orders");
  System.out.println(order);

  assertThat(order).hasNoNullFieldsOrProperties();
}
}
```

这里有两种测试方法：普通方法将消息手动转换为 Order 对象，而另一个方法使用 receiveAndConvert 方法来完成转换。两种测试方法进行比较，可以看出 MessageConverter 完成了什么工作，同时能发现它使得代码更易读。MappingJackson2MessageConverter 将 Order 对象转换为 BytesMessages。要使用 TextMessage，可以将 targetType 属性设置为 TEXT。然后，应用程序将接收一条 String 类型的消息，其有效载荷是一个 JSON 对象。

9.3 使用 JMS 接收消息

9.3.1 问题

你希望从某个 JMS 目的地读取消息,以便在应用程序中处理这些消息。

9.3.2 解决方案

创建一个类并使用@JmsListener 对这个类的方法进行注解,以将其绑定到某个目的地并处理传入消息。

9.3.3 工作原理

可以创建一个 POJO(Plain Ordinary Java Object,简单的 Java 对象)并使用@JmsListener 对其方法进行注解。Spring 将检测到这个对象,并为其创建一个 JMS 侦听器。Spring Boot 在 spring.jms.listener 名称空间中公开了一些属性,用于配置这个侦听器。

1. 接收消息

让我们创建一个服务来监听 9.2 小节的发送者程序发送的消息。

```
@Component
class CurrentDateTimeService {

  @JmsListener(destination = "time-queue")
  public void handle(Message msg) throws JMSException {
    Assert.state(msg instanceof TextMessage, "Can only handle
    TextMessage.");
    System.out.println("[RECEIVED]-"+((TextMessage) msg).getText());
  }
}
```

上述代码是一个常规类,有一个用@JmsListener 注解的方法,该方法至少需要一个目的地(通过 destination 属性设置),以便知道从何处检索消息。该方法接收 javax.jms.Message 类型的参数;我们在代码中限制只处理 TextMessage 类型的消息,并将内容打印到控制台。@JmsListener 注解的方法的签名比较灵活,因为它允许多个注解的或特定类型(如表 9-6 所示)的参数。

表 9-6　可以使用的方法参数类型

类型	说明
java.lang.String	仅用于 TextMessage 消息类型，将消息的有效载荷作为 String 类型的数据
java.util.Map	仅用于 MapMessage 消息类型，将消息的有效载荷作为 Map 类型的数据
byte[]	仅用于 BytesMessage 消息类型，将消息的有效载荷作为 byte[]类型的数据
Serializable object	从 ObjectMessage 反序列化 Object
javax.jms.Message	获取具体的 JMS 消息
javax.jms.Session	为获取访问读取 Session，例如发送自定义响应消息
@Header annotated element	从 JMS 消息提取消息头
@Headers annotated element	仅用于 java.util.Map 消息类型，从 JMS 消息中提取所有的消息头

可以通过简单地使用 String 对象作为方法参数来简化侦听器，而不用处理 javax.jms.Message 消息类型的参数。

```
@Component
class CurrentDateTimeService {

  @JmsListener(destination = "time-queue")
  public void handle(String msg) {
        System.out.println("[RECEIVED] - " + msg);
  }
}
```

2. 配置侦听器容器

Spring 使用 JmsListenerContainerFactory 创建支持@JmsListener 注解所需的基础架构。Spring Boot 配置了一个默认 JmsListenerContainerFactory 实例，可以使用 spring.jms.listener 名称空间中的属性对其进行配置(见表 9-7)。如果这还不能满足要求，你完全可以配置自己的实例，并手动配置所有配置选项。此时，当 Spring Boot 在上下文中检测到 JmsListenerContainerFactory 实例时，它将避免创建新的 JmsListenerContainerFactory 实例。

表 9-7 侦听器容器的配置属性

属性	说明
spring.jms.listener.acknowledge-mode	容器的确认模式，默认为自动
spring.jms.listener.auto-startup	系统启动时自动启动容器，默认为 true
spring.jms.listener.concurrency	并发消费者的最小数量，默认值为空，表示 1 个并发使用者(Spring 的默认值)
spring.jms.listener.max-concurrency	并发消费者的最大数目，默认值为空，表示 1 个并发使用者(Spring 默认值)
spring.jms.pub-sub-domain	默认目标是主题。默认值为 false，表示默认目标是队列

默认配置的 JmsListenerContainerFactory 还将检测是否存在单个独特的 DestinationResolver 和 MessageConverter，如果找到，将使用这两个实例；否则，它将使用 Spring 默认的 DynamicDestinationResolver 和 SimpleMessageConverter 实例(更多信息请参阅 9.2 小节)。

3. 使用自定义 MessageConverter

如果想要通过 JMS 将 Java 对象作为 JSON 对象发送到下一个系统，该怎么处理？你可以使用 Java 序列化，但通常不赞成这样做，因为这种方式和系统的耦合度太紧密。使用 JSON 或 XML 传输对象/消息是更好的方法。使用 Spring JMS，这只需要配置不同的 MessageConverter(另请参见 9.2 小节中发送消息相关的内容)即可。

```
@Bean
public MappingJackson2MessageConverter messageConverter() {
    var messageConverter = new MappingJackson2MessageConverter();
    messageConverter.setTypeIdPropertyName("content-type");
    messageConverter.setTypeIdMappings(singletonMap("order",
      Order.class));
    return messageConverter;
}
```

默认情况下，MappingJackson2MessageConverter 要求在内容类型的标识符中放置一个属性名称。这将从 JMS 消息的消息头中读取(这里我们将其设置为 content-type 的值)。接下来，我们可以选择定义类型和类之间的映射。因为我们希望能够将 order 对象映射到 Order 类，所以我们将 Order.class 指定为 content-type 中 order 的映射。

```
@Component
class OrderService {
```

```
    @JmsListener(destination = "orders")
    public void handle(Order order) {
        System.out.println("[RECEIVED] - " + order);
    }
}
```

侦听器接收 Order 对象,而 Spring JMS 将负责接收和转换消息。如果将这个侦听器与 9.2 小节中的订单发送器组合在一起,你将看到一个稳定的订单流。

4. 发送回复

有时,当接收到消息时,希望返回一个响应或触发流程的另一部分。使用 Spring 消息传递功能实现这一点非常简单:直接从处理器方法返回想要发送的内容。为了确定将响应发送到何处,可以添加附加的@SendTo 注解以指定目标。让我们修改示例,将 OrderConfirmation 对象发送到 order-confirmations 队列。

```
@Component
class OrderService {

    @JmsListener(destination = "orders")
    @SendTo("order-confirmations")
    public OrderConfirmation handle(Order order) {

        System.out.println("[RECEIVED] - " + order);
        return new OrderConfirmation(order.getId());
    }
}
```

OrderService 类稍有变化,现在在处理订单之后返回一个 OrderConfirmation 对象。通过@SendTo 注解,我们指定将结果返回到哪个目的地。

接下来针对 OrderConfirmation 对象创建另一个侦听器,以便接收这些响应。

```
@Component
class OrderConfirmationService {

    @JmsListener(destination = "order-confirmations")
    public void handle(OrderConfirmation confirmation) {
        System.out.println("[RECEIVED] - " + confirmation);
    }
}
```

最后，创建 OrderConfirmation 类，根据传入的消息构造类的实例。

```java
public class OrderConfirmation {

  private String orderId;

  public OrderConfirmation() {}

  public OrderConfirmation(String orderId) {
     this.orderId = orderId;
  }

  public String getOrderId() {
     return orderId;
  }

  public void setOrderId(String orderId) {
    this.orderId = orderId;
  }

  @Override
  public String toString() {
    return String.format("OrderConfirmation [orderId='%s']",
      orderId);
  }
}
```

运行应用程序时将看到，首先接收 Order 消息，然后接收 OrderConfirmation 消息。

9.4 配置 RabbitMQ

9.4.1 问题

你希望在 Spring Boot 应用程序中使用 AMQP 消息传递功能，并且需要连接到 RabbitMQ 代理。

9.4.2 解决方案

配置适当的 spring.rabbitmq 属性(最少需要配置 spring.rabbitmq.host 属性)以连接

到交换,从而能够发送和接收消息。

9.4.3 工作原理

当 Spring Boot 检测到类路径上的 RabbitMQ 客户端库时,它将自动创建一个 ConnectionFactory 实例。作为开始,需要添加 spring-boot-starter-amqp 依赖项,这将引入所有必需的依赖关系。

```
<dependency>
  <groupId>org.springframework.boot</groupId>
  <artifactId>spring-boot-starter-amqp</artifactId>
</dependency>
```

现在可以通过配置 spring.rabbitmq 属性连接到 RabbitMQ 代理,如下所示。

```
spring.rabbitmq.host=localhost
spring.rabbitmq.port=5672
spring.rabbitmq.username=guest
spring.rabbitmq.password=guest
```

上述配置是连接到默认 RabbitMQ 实例所需的全部内容,也是 Spring Boot 所使用的默认配置,见表 9-8。

表 9-8 常用的 RabbitMQ 配置属性

属性	说明
spring.rabbitmq.addresses	客户机应连接到的地址列表,多个地址之间以逗号分隔
spring.rabbitmq.connection-timeout	连接超时时间。默认为空,0 表示从不超时
spring.rabbitmq.host	RabbitMQ 主机,默认为本地主机(localhost)
spring.rabbitmq.port	RabbitMQ 端口,默认为 5672
spring.rabbitmq.username	连接 RabbitMQ 代理使用的用户名,默认为 guest
spring.rabbitmq.password	连接 RabbitMQ 代理使用的用户密码,默认为 guest
spring.rabbitmq.virtual-host	连接到代理时使用的虚拟主机

9.5 使用 RabbitMQ 发送消息

9.5.1 问题

你希望将消息发送给一个 RabbitMQ 代理,以便将消息传递到接收方。

9.5.2 解决方案

使用 RabbitTemplate，你可以将消息发送到交换机并提供路由键。

9.5.3 工作原理

Spring Boot 在检测到唯一的 ConnectionFactory 实例时自动配置一个 RabbitTemplate 实例，这个模板实例可用于向队列发送消息。

1. 配置 RabbitTemplate

如果 Spring Boot 可以检测到唯一的 ConnectionFactory 实例，并且配置中没有 RabbitTemplate 实例存在，那么它将自动配置一个 RabbitTemplate 实例。Spring Boot 允许通过 spring.rabbitmq.template 名称空间中的属性(如表 9-9 所示)修改已配置好的 RabbitTemplate 实例。

表 9-9 RabbitTemplate 配置属性

属性	说明
spring.rabbitmq.template.exchange	用于发送操作的默认交换机的名称，默认为空
spring.rabbitmq.template.routing-key	用于发送操作的默认路由键的键值，默认为空
spring.rabbitmq.template.receive-timeout	接收操作的超时时间，默认为 0，表示永不超时
spring.rabbitmq.template.reply-timeout	发送-接收操作的超时时间，默认为 5 秒钟

Spring Boot 还可以很容易地配置 RabbitTemplate 实例的重试逻辑，默认情况下该特性是禁用的。通过在 application.properties 文件中添加配置 spring.rabbitmq.template.retry.enabled=true 启用该特性。现在，当发送失败时，应用程序将另外尝试两次发送消息。如果要修改重试次数或间隔，可以使用 spring.rabbitmq.template.retry 名称空间中的属性，如表 9-10 所示。

表 9-10 RabbitTemplate 重试配置属性

属性	说明
spring.rabbitmq.template.retry.enabled	启用重试功能，默认为 false
spring.rabbitmq.template.retry.max-attempts	尝试发送消息的次数，默认为 3 次
spring.rabbitmq.template.retry.initial-interval	第一次和第二次尝试发送消息之间的间隔，默认为 1 秒
spring.rabbitmq.template.retry.max-interval	两次尝试发送消息的最大时间间隔，默认为 10 秒
spring.rabbitmq.template.retry.multiplier	应用于前一个间隔的乘数，默认值为 1.0

2. 发送简单的消息

使用 RabbitTemplate 发送消息可以通过调用 convertAndSend 方法来完成。该方法至少需要路由键和在要发送的消息中包含的对象作为参数。

```
@Component
class HelloWorldSender {

  private final RabbitTemplate rabbit;

  HelloWorldSender(RabbitTemplate rabbit) {
     this.rabbit = rabbit;
  }

  @Scheduled(fixedRate = 500)
  public void sendTime() {
     rabbit.convertAndSend("hello",
                   "Hello World, from Spring Boot 2, over
                   RabbitMQ!");
  }
}
```

HelloWorldSender 将通过构造函数注入 RabbitTemplate 实例。该实例通过 hello 路由键每隔 500 毫秒向默认交换机发送一条消息。当消息发送到默认交换机时，默认交换机将自动创建一个名为 hello 的队列。你可以在 RabbitMQ 管理控制台(默认地址是 http://localhost:15672)中查看队列中的消息数量。

下面编写一个测试来验证应用程序的行为是否正确。由于 RabbitMQ 没有嵌入式代理，因此需要使用@MockBean 模拟 RabbitTemplate。在@Test 方法中，使用适当的参数验证方法调用。

```
@RunWith(SpringRunner.class)
@SpringBootTest
public class RabbitSenderApplicationTest {

  @MockBean
  private RabbitTemplate rabbitTemplate;

  @Test
  public void shouldSendAtLeastASingleMessage() {
     verify(rabbitTemplate, Mockito.atLeastOnce())
```

```
        .convertAndSend("hello",
                "Hello World, from Spring Boot 2, over RabbitMQ!");
    }
}
```

> **提示：**
> 可以通过多种方法使用Docker或嵌入式进程[3]来启动RabbitMQ。上述代码编写的测试中就包括了一个嵌入式进程。

3. 发送对象

为了向 RabbitMQ 发送消息，消息有效载荷必须转换为一个 byte[]数组。对于一个 String 类型的对象，通过调用 String.getBytes 即可完成转换。然而，当发送其他类型的对象时，转换过程变得比较麻烦。默认实现将检查该对象是否可序列化，如果是，将使用 Java 序列化将对象转换为一个 byte[]数组。使用 Java 序列化并不是最好的解决方案，尤其是当你需要向非 Java 客户机发送消息时。

RabbitTemplate 使用 MessageConverter 将消息的创建委托给特定的类。默认情况下，MessageConverter 使用 SimpleMessageConverter 类，该类实现了上述粗略描述的策略。同时，对于以 XML 或 JSON 作为有效载荷的消息，Spring Boot 提供了不同的实现来完成转换。对于 XML，由 MarshallingMessageConverter 类完成转换；对于 JSON，由 Jackson2JsonMessageConverter 类完成转换。

Spring Boot 将自动检测已配置的 MessageConverter，并将其用于 RabbitTemplate 和侦听器(参见 9.5 小节)。

```
@Bean
public Jackson2JsonMessageConverter jsonMessageConverter() {
  return new Jackson2JsonMessageConverter();
}
```

这足以将 SimpleMessageConverter 对象转变为 Jackson2JsonMessageConverter 对象。

让我们创建一个 Order 对象，并使用 RabbitTemplate 以 new-order 作为路由键将其发送到名为 orders 的交换机。

```
package com.apress.springbootrecipes.demo;

import java.math.BigDecimal;

public class Order {
```

---

[3] https://github.com/AlejandroRivera/embedded-rabbitmq.

```
    private String id;
    private BigDecimal amount;

    public Order() {
    }

    public Order(String id, BigDecimal amount) {
        this.id=id;
        this.amount = amount;
    }

    public String getId() {
        return id;
    }

    public void setId(String id) {
        this.id = id;
    }

    public BigDecimal getAmount() {
        return amount;
    }

    public void setAmount(BigDecimal amount) {
      this.amount = amount;
    }

    @Override
    public String toString() {
        return String.format("Order[id='%s',amount=%4.2f]",id,amount);
    }
}
```

Order 类已创建好，接下来创建一个方法，该方法周期性地发送随机生成的 order 对象。

```
@Component
class OrderSender {
```

```java
    private final RabbitTemplate rabbit;

    OrderSender(RabbitTemplate rabbit) {
        this.rabbit = rabbit;
    }

    @Scheduled(fixedRate = 256)
    public void sendTime() {
      var id = UUID.randomUUID().toString();
      var amount = ThreadLocalRandom.current().nextDouble(1000.00d);
      var order = new Order(id, BigDecimal.valueOf(amount));
      rabbit.convertAndSend("orders", "new-order", order);
    }
}
```

上述代码将创建最大数量为 1000 的随机数量的订单。然后，它将使用 convertAndSend 方法，以 new-order 作为路由键将消息发送到名为 orders 的交换机。

测试中将再次使用@MockBean 创建 RabbitTemplate 的模拟实例，并测试方法调用。

```java
@RunWith(SpringRunner.class)
@SpringBootTest
public class RabbitSenderApplicationTest {

  @MockBean
  private RabbitTemplate rabbitTemplate;

  @Test
  public void shouldSendAtLeastASingleMessage() {

      verify(rabbitTemplate, atLeastOnce())
            .convertAndSend(
                    eq("orders"),
                    eq("new-order"),
                    any(Order.class));
  }
}
```

4. 编写集成测试

为嵌入式 RabbitMQ 服务器添加依赖项，以便轻松地在测试用例中启动它。

```xml
<dependency>
  <groupId>io.arivera.oss</groupId>
  <artifactId>embedded-rabbitmq</artifactId>
 <version>1.3.0</version>
  <scope>test</scope>
</dependency>
```

接下来，创建 RabbitSenderApplicationIntegrationTestConfiguration 类，其中包含运行集成测试所需的附加配置。

```java
@TestConfiguration
public class RabbitSenderApplicationIntegrationTestConfiguration {

  @Bean(initMethod = "start", destroyMethod = "stop")
  public EmbeddedRabbitMq embeddedRabbitMq() {
    EmbeddedRabbitMqConfig config = new
        EmbeddedRabbitMqConfig.Builder()
           .rabbitMqServerInitializationTimeoutInMillis(10000).
             build();
   return new EmbeddedRabbitMq(config);
  }

    @Bean
   public Queue newOrderQueue() {
       return QueueBuilder.durable("new-order").build();
   }

    @Bean
   public Exchange ordersExchange() {
   return ExchangeBuilder.topicExchange("orders").durable(true).build();
   }

    @Bean
   public Binding newOrderQueueBinding(Queue queue, Exchange exchange) {
     return BindingBuilder.bind(queue).to(exchange)
                    .with("new-order").noargs();
```

    }
}
```

配置内容包括嵌入式 RabbitMQ 定义、Queue 和 Exchange 定义,最后使用路由键 new-order 将 Queue 绑定到 Exchange。

为了接收消息,需要创建队列并进行绑定,否则队列只会驻留在交换上(或者,根据配置,将被丢弃)。

集成测试将加载应用程序和附加配置,它将使用 RabbitTemplate 接收消息。

```java
@RunWith(SpringRunner.class)
@SpringBootTest(classes = {
        RabbitSenderApplication.class,
        RabbitSenderApplicationIntegrationTestConfiguration.class })
public class RabbitSenderApplicationIntegrationTest {

    @Autowired
    private RabbitTemplate rabbitTemplate;

    @Test
    public void shouldSendAtLeastASingleMessage() {

        Message msg = rabbitTemplate.receive("new-order", 1500);

        assertThat(msg).isNotNull();
        assertThat(msg.getBody()).isNotEmpty();
        assertThat(msg.getMessageProperties().getReceivedExchange())
                .isEqualTo("orders");
        assertThat(msg.getMessageProperties().getReceivedRoutingKey())
                .isEqualTo("new-order")
        assertThat(msg.getMessageProperties().getContentType())
                .isEqualTo(MediaType.APPLICATION_JSON_VALUE);
    }
}
```

如上所述,测试将加载应用程序和其他配置类。这是通过设置 @SpringBootApplication 注解中的 classes 属性实现的。当测试开始时,它将从配置中定义的 new-order 队列接收一条消息。然后使用断言对接收到的消息进行验证,验证的内容包括编码(application/json)、路由键等。
```

9.6 使用 RabbitMQ 接收消息

9.6.1 问题

你希望接收来自 RabbitMQ 的消息。

9.6.2 解决方案

用@RabbitListener 注解一个方法会将其绑定到一个队列,并让它接收队列上的消息。

9.6.3 工作原理

如果 bean 的方法使用了@RabbitListener 注解,这个 bean 将用作传入消息的消息侦听器。Spring Boot 在检测到@RabbitListener 注解时会构造一个消息侦听器容器,带@RabbitListener 注解的方法将接收传入的消息。消息侦听器容器可以通过 spring.rabbitmq.listener 名称空间中的属性进行配置,见表 9-11。

表 9-11 Rabbit Listener 属性

属性	说明
spring.rabbitmq.listener.type	侦听器容器的类型,包括 direct 和 simple 类型,默认为 simple
spring.rabbitmq.listener.simple.acknowledge-mode	容器确认模式,默认为空
spring.rabbitmq.listener.simple.prefetch	单个请求中可以处理的消息数量,默认为空
spring.rabbitmq.listener.simple.default-requeue-rejected	是否将被拒绝的消息重新放入队列中
spring.rabbitmq.listener.simple.concurrency	侦听器触发器线程的最小数量
spring.rabbitmq.listener.simple.max-concurrency	侦听器触发器线程的最大数量
spring.rabbitmq.listener.simple.transaction-size	单个事务中处理的消息数量。为了获得最佳结果,该属性的值应小于等于 prefetch 属性的值
spring.rabbitmq.listener.direct.acknowledge-mode	容器确认模式,默认为空
spring.rabbitmq.listener.direct.prefetch	单个请求中可以处理的消息数量,默认为空
spring.rabbitmq.listener.direct.default-requeue-rejected	是否将被拒绝的消息重新放入队列中
spring.rabbitmq.listener.direct.consumers-per-queue	每个队列的消费者数量,默认值为 1

1. 接收简单的消息

要开始接收来自 RabbitMQ 的消息,只需要一个带有@RabbitListener 注解的组件即可。

```
@Component
class HelloWorldReceiver {

  @RabbitListener( queues = "hello")
  public void receive(String msg) {
      System.out.println("Received: " + msg);
  }
}
```

上述组件将接收来自 hello 队列的所有消息,并将其打印到控制台。对于简单的有效负载或者如果接收对象可以从消息的有效负载反序列化,该组件就足以完成接收消息的任务。但是,在发送对象或复杂消息时,人们可能更喜欢使用 JSON 或 XML。

2. 接收对象

若要在不依赖 Java 序列化的情况下接收更复杂的对象,则需要配置一个 MessageConverter 实例(请参见 9.4 小节)。消息侦听器容器将使用已配置的 MessageConverter 将传入的有效负载转换为@RabbitListener 注解的方法所需的对象。

```
@Bean
public Jackson2JsonMessageConverter jsonMessageConverter() {
  return new Jackson2JsonMessageConverter();
}
```

为了配置 MessageConverter,需要创建一个@Bean 注解的方法并构造希望使用的转换器。这里构造的是基于 Jackson 2 的转换器,但是也有一个用于解组 XML 的转换器,即 MarshallingMessageConverter。

```
@Component
class OrderService {

  @RabbitListener(bindings = @QueueBinding(
      exchange=@Exchange(name="orders", type = ExchangeTypes.TOPIC),
      value = @Queue(name = "incoming-orders"),
      key = "new-order"
  ))
  public void handle(Order order) {
```

```
    System.out.println("[RECEIVED] - " + order);
  }
}
```

上述侦听器将使用 orders 交换机(这是一个主题类型的交换机)，并使用 new-order 路由键为 incoming-orders 队列创建绑定。应用程序启动时，如果交换机和队列不存在，将自动创建它们。传入的消息使用 Jackson2JsonMessageConverter 转换为 order 对象。

3. 接收消息并发送响应

当接收到消息时，可能需要将响应发送回客户端或通过不同的消息进行通信。可以在返回类型为非 void 的方法上使用@RabbitListener；它将创建一个结果消息，并将该消息放在带有路由键的交换机上；发送消息的目的地需要在@SendTo 注解中指定。

```
@Component
class OrderService {

  @RabbitListener(bindings = @QueueBinding(
     exchange=@Exchange(name="orders", type = ExchangeTypes.TOPIC),
     value = @Queue(name = "incoming-orders"),
     key = "new-order"

  ))
  @SendTo("orders/order-confirmation")
  public OrderConfirmation handle(Order order) {
     System.out.println("[RECEIVED] - " + order);
     return new OrderConfirmation(order.getId());
  }
}
```

应用程序接收并处理完 Order 消息时，将发送 OrderConfirmation 消息，@SendTo 注解(来自通用的 Spring Messaging 组件)包含交换机和路由键。在斜线之前的部分是交换机，之后的部分是路由键，因此使用的模式是<exchange>/<routing-key>。交换机或路由键(或两者同时)可以为空，在这种情况下，将使用默认配置的交换机和路由键。在这里，代码使用 orders 交换机并使用 order-confirmation 作为路由键。

可以使用另一个侦听器来处理 OrderConfirmation 消息。

```
@Component
class OrderConfirmationService {
```

```
@RabbitListener(bindings = @QueueBinding(
    exchange=@Exchange(name="orders", type = ExchangeTypes.TOPIC),
    value = @Queue(name = "order-confirmations"),
    key = "order-confirmation"
))

public void handle(OrderConfirmation confirmation) {
    System.out.println("[RECEIVED] - " + confirmation);
}
}
```

上述代码将使用 order-confirmation 路由键创建 order-confirmations 队列,并在 orders 交换机上绑定该路由键(就像前面创建的 OrderService)。在与 9.4 小节的发送者程序一起运行上述代码时,程序应该会收到 order 对象,并发送对应的确认消息。

第 10 章

Spring Boot Actuator

在开发应用程序时，你还希望能够监视应用程序的行为。Spring Boot 通过引入 Spring Boot Actuator 使得实现这个特性变得很容易。Spring Boot Actuator 将应用程序中的健康状况和性能指标公开给相关方。这些数据可以通过 JMX、HTTP 传输，或者导出到一个外部系统。

健康状况端点显示应用程序和/或运行它的系统的健康状况。它将检测数据库是否已启动、报告磁盘空间等。性能指标端点会公开使用情况和性能统计信息，如请求数、最长的请求、最快的请求、连接池的使用率等。

所有这些指标都可以在启用时通过 JMX 或 HTTP 查看，但也可以自动导出到外部系统，如 Graphite、InfluxDB 和许多其他系统。

10.1 启用和配置 Spring Boot Actuator

10.1.1 问题

你希望在应用程序中启用健康和性能监控，这样就可以监视应用程序的状态。

10.1.2 解决方案

在项目中添加 spring-boot-starter-actuator 依赖项，Spring Boot 将启动健康和性能监控，并将相关数据向外部公开。可以通过 management 名称空间中的属性进行其他配置。

10.1.3 工作原理

在添加了 spring-boot-starter-actuator 依赖项之后，Spring Boot 将根据应用程序上下文中的 bean 自动设置收集和公开哪些健康状况和性能指标数据。哪些数据被公开取决于存在哪些 bean 和启用的特性。在检测到 DataSource 时，将收集和公开该数据源的性能指标以及健康状况的数据。Spring Boot 可以为许多组件执行此操作，如 Hibernate、RabbitMQ、Caches 等。

为了启用 Spring Boot Actuator，必须在应用程序中添加上述依赖项(在这里我们以 3.3 小节的源代码作为起点)。

```
<dependency>
  <groupId>org.springframework.boot</groupId>
  <artifactId>spring-boot-starter-actuator</artifactId>
</dependency>
```

现在在启动应用程序时，Spring Boot 将配置好 Spring Boot Actuator，并可以通过 JMX(如图 10-1 所示，关于如何使用 JConsole 请参见 8.4 小节)和网络(如图 10-2 所示，默认路径是/actuator)进行访问。

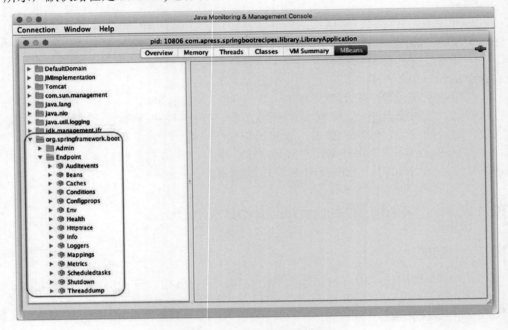

图 10-1　JMX 公开的性能指标

可以看到，JMX 比 HTTP 公开更多的端点。HTTP 只公开/actuator/health 和 /actuator/info。JMX 公开了更多。这是出于安全考虑：/actuator 是公开的，而你不希望每个人都能看到应用程序的运行情况。可以通过 management.endpoints.web.exposure.include 和 management.endpoints.web.exposure.exclude 属性配置要公开的内容。在 include 属性中使用*将向 Web 公开所有端点。

```
management.endpoints.web.exposure.include=*
```

在 application.properties 文件中添加上述配置之后，通过 HTTP 访问将和通过 JMX 获取的数据一样，如图 10-3 所示。

第 10 章 Spring Boot Actuator

```
{
    - _links: {
        - self: {
            href: "http://localhost:8080/actuator",
            templated: false
        },
        - health-component: {
            href: "http://localhost:8080/actuator/health/{component}",
            templated: true
        },
        - health-component-instance: {
            href: "http://localhost:8080/actuator/health/{component}/{instance}",
            templated: true
        },
        - health: {
            href: "http://localhost:8080/actuator/health",
            templated: false
        },
        - info: {
            href: "http://localhost:8080/actuator/info",
            templated: false
        }
    }
}
```

图 10-2　HTTP 访问公开的性能指标

```
{
    - _links: {
        - self: {
            href: "http://localhost:8080/actuator",
            templated: false
        },
        - auditevents: {
            href: "http://localhost:8080/actuator/auditevents",
            templated: false
        },
        - beans: {
            href: "http://localhost:8080/actuator/beans",
            templated: false
        },
        - caches-cache: {
            href: "http://localhost:8080/actuator/caches/{cache}",
            templated: true
        },
        - caches: {
            href: "http://localhost:8080/actuator/caches",
            templated: false
        },
        - health: {
            href: "http://localhost:8080/actuator/health",
            templated: false
        },
        - health-component: {
            href: "http://localhost:8080/actuator/health/{component}",
            templated: true
        },
        - health-component-instance: {
            href: "http://localhost:8080/actuator/health/{component}/{instance}",
            templated: true
        },
```

图 10-3　HTTP 访问获取所有性能指标

1. 配置管理服务器

默认情况下，访问 Spring Boot Actuator 的地址和端口与应用程序的常规地址和端口(http://localhost:8080)相同。然而，可以很轻松地配置在不同的端口上运行管理端点，这也是普遍的做法。这可以通过 management.server 名称空间中的属性进行配置，这些属性大多数都模仿常规 server 名称空间中的属性，如表 10-1 所示。

表 10-1　管理服务器的属性

属性	说明
management.server.add-application-context-header	将 X-Application-Context 头添加到包含应用程序上下文名称的响应中
management.server.port	运行管理服务器的端口号，默认与 server.port 属性的值相同
management.server.address	要绑定到的网络地址，默认与 server.address 属性的值相同(即 0.0.0.0，表示所有地址)
management.server.servlet.context-path	管理服务器的上下文路径，默认为空，表示根目录/
management.server.ssl.*	用于配置管理服务器的 SSL 的属性(有关如何配置 SSL 的信息，请参阅 3.8 小节)

将下面的配置添加到 application.properties 文件中，Spring Boot 将在一个单独的端口上运行管理端点，并添加 X-Application-Context 头。

```
management.server.add-application-context-header=true
management.server.port=8090
```

重新启动应用程序，现在可以在 http://localhost:8090/actuator 上访问管理端点。在不同端口上运行 Spring Boot Actuator 的好处是，通过在防火墙上封锁该端口，只允许本地访问，从而将其隐藏于公共互联网。

■ 注意：
management.server 名称空间中的属性只在使用嵌入式服务器时才有效，当部署到外部服务器时，这些属性就不再适用了！

2. 配置单个管理端点

可以通过 management.endpoint.<endpoint-name>名称空间中的属性来配置各个端点。其中大多数端点至少有 enabled 属性和 cache.time-to-live 属性。第一个属性将启用或禁用端点，另一个指定缓存端点的结果的时间长度，如表 10-2 所示。

表 10-2　端点配置属性

属性	说明
management.endpoint.<endpoint-name>.enabled	是否启用特定端点，通常默认为 true，有时还取决于功能的可用性(例如，如果没有启用 Flyway，那么启用 flyway 端点就不会有任何效果)
management.endpoint.<endpoint-name>.cache.time-to-live	缓存响应的时间长度，默认为 0ms，表示不缓存
management.endpoint.health.show-details	是否显示健康端点的详细信息，默认为 never，可以设置为 always 或 when-authorized
management.endpoint.health.roles	允许查看消息信息的角色(当 show-details 属性设置为 when-authorized 时配置该属性)

将 management.endpoint.health.show-details=always 添加到 application.properties 文件中，会显示更多关于应用程序健康状况的信息。默认情况下，它将只显示应用程序的状态为"UP"，但现在可以看到应用程序的更多健康信息，如图 10-4 所示。

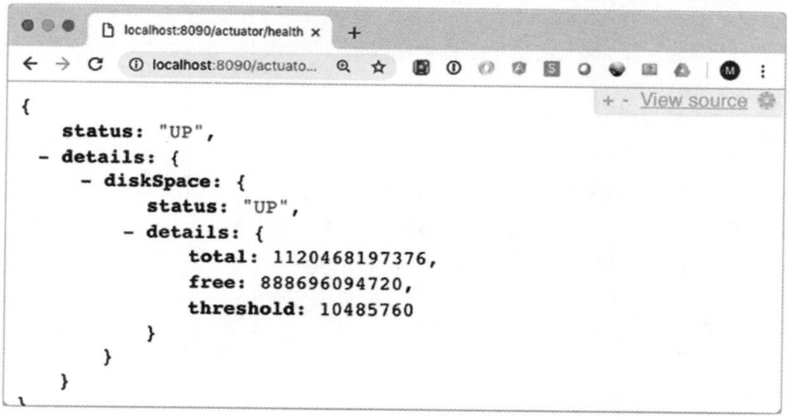

图 10-4　健康状况端点的扩展输出

3. 管理端点的安全保护

当 Spring Boot 检测到 Spring Boot Actuator 和 Spring Security 时，它将自动启用对管理端点的安全保护。访问端点时，界面将显示基本登录提示，并要求输入用户名和密码。Spring Boot 将生成一个默认用户用于登录，默认用户的用户名是 user，密码是系统生成的(参见 6.1 小节)。

除了添加 spring-boot-starter-actuator 之外，再添加 spring-boot-starter-security 就足以保护管理端点。

```
<dependency>
    <groupId>org.springframework.boot</groupId>
```

```
<artifactId>spring-boot-starter-security</artifactId>
</dependency>
```

这将为应用程序和管理端点启用安全特性。现在，访问端点 http://localhost:8090/actuator 时，将显示基本登录提示。在输入正确的凭据之后，你应该仍然能够看到结果。

4. 配置健康状况检查

Spring Boot Actuator 的一个特性是进行健康状况检查，应用程序的健康状况数据可以访问 http://localhost:8090/actuator/health 获取。首先看到的信息是应用程序是运行状态(UP)还是关闭状态(DOWN)。健康端点调用系统中所有可用的 HealthIndicator，并在端点中展现收集到的数据。可以通过设置 management.health.<health-indicator>.enabled 属性来控制展现哪些 HealthIndicator 的数据。将不可用特性的属性设置为 true(例如在 DataSource 不可用时尝试获取其信息)是无法收集到相关数据的。

```
management.health.diskspace.enabled=false
```

上述代码将禁止查看磁盘空间的健康数据，该数据将不会再出现在健康状况检查报告中。

5. 配置性能指标

Spring Boot Actuator 的一个特性是公开性能指标。应用程序的性能指标数据可以访问 http://localhost:8090/actuator/metrics 获取，浏览器将展现应用程序可用的性能指标列表，如图 10-5 所示。

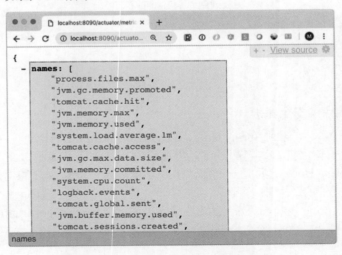

图 10-5　当前可用的性能指标列表

访问 http://localhost:8090/actuator/metrics/{name-of-metric}可以获得有关性能指标的更多信息，例如 http://localhost:8090/actuator/metrics/system.cpu.usage 将显示当前的 CPU 使用情况，如图 10-6 所示。

```
{
    name: "system.cpu.usage",
    description: "The "recent cpu usage" for the whole system",
    baseUnit: null,
  - measurements: [
      - {
            statistic: "VALUE",
            value: 0.4375
        }
    ],
    availableTags: [ ]
}
```

图 10-6　详细的 CPU 性能指标

Spring Boot 使用 micrometer.io[1] 来记录性能指标。对于检测到的特性，Spring Boot 默认启动记录性能指标。因此，如果检查到 DataSource，Spring Boot 将记录相关的性能指标。要禁用指标，需要添加 management.metrics.enable.*属性，其中*代表需要配置的指标。这种配置方式包含键值对的映射，*的内容将被映射到对应的属性上。

```
management.metrics.enable.system=false
management.metrics.enable.tomcat=false
```

上述配置将禁止记录 system 和 tomcat 性能指标。在访问 http://localhost:8090/actuator/metrics 查看 Spring Boot 记录的性能指标时，将不会看到这两个指标的数据。

10.2　创建自定义的健康状况检查和性能指标

10.2.1　问题

应用程序需要公开某些指标，并检查默认情况下不可用的健康状况数据。

---

[1] https://micrometer.io.

10.2.2 解决方案

健康检查和性能指标是可插拔的,类型为 HealthIndicator 和 MetricBinder 的 bean 将自动注册,以提供额外的健康状况检查和/或性能指标。我们的任务是创建一个类,该类实现所需的接口并将该类的一个实例注册为 bean,使其能够提供所需的健康状况和性能指标数据。

10.2.3 工作原理

假设已经添加 Spring Boot Actuator 的依赖项,现在可以立即开始编写一个实现。假设你有一个使用 TaskScheduler 的应用程序,并且希望记录该应用程序的一些性能指标和健康状况。为了让 Spring Boot 默认创建一个 TaskScheduler 实例,只需要在应用程序的类上添加@EnableScheduling 注解即可。

首先,让我们编写 HealthIndicator。你可以直接实现 HealthIndicator 接口,也可以使用 AbstractHealthIndicator 作为基类。

```java
package com.apress.springbootrecipes.library.actuator;

import org.springframework.boot.actuate.health.AbstractHealthIndicator;
import org.springframework.boot.actuate.health.Health;
import org.springframework.scheduling.concurrent.ThreadPoolTaskScheduler;
import org.springframework.stereotype.Component;

@Component
class TaskSchedulerHealthIndicator extends AbstractHealthIndicator {
  private final ThreadPoolTaskScheduler taskScheduler;

  TaskSchedulerHealthIndicator(ThreadPoolTaskScheduler taskScheduler) {
    this.taskScheduler = taskScheduler;
  }

  @Override
  protected void doHealthCheck(Health.Builder builder) throws Exception {

    int poolSize = taskScheduler.getPoolSize();
    int active = taskScheduler.getActiveCount();
```

```
        int free = poolSize - active;

        builder
            .withDetail("active", taskScheduler.getActiveCount())
            .withDetail("poolsize", taskScheduler.getPoolSize());

        if (poolSize > 0 && free <= 1) {
            builder.down();
      } else {
          builder.up();
        }
      }
    }
```

TaskSchedulerHealthIndicator 类在特定的 ThreadPoolTaskExecutor 实例上进行操作。如果只有一个或者很少的线程可用于调度任务，则将报告状态为"DOWN"。之所以使用 poolSize>0 作为判断条件，是因为只有在需要时才会创建底层的 Executor；在这之前，poolSize 将显示为 0。返回的值包括 poolsize 和活跃的线程数，仅作为参考信息。

TaskSchedulerMetrics 类实现了 micrometer.io 包中的 MeterBinder 接口。它向性能指标注册表公开 active 和 pool-size 指标，如下所示。

```
package com.apress.springbootrecipes.library.actuator;

import io.micrometer.core.instrument.FunctionCounter;
import io.micrometer.core.instrument.MeterRegistry;
import io.micrometer.core.instrument.binder.MeterBinder;
import org.springframework.scheduling.concurrent
.ThreadPoolTaskScheduler;
import org.springframework.stereotype.Component;
@Component
class TaskSchedulerMetrics implements MeterBinder {

  private final ThreadPoolTaskScheduler taskScheduler;

  TaskSchedulerMetrics(ThreadPoolTaskScheduler taskScheduler) {
      this.taskScheduler = taskScheduler;
  }
```

```
@Override
public void bindTo(MeterRegistry registry) {
    FunctionCounter
            .builder("task.scheduler.active", taskScheduler,
                    ThreadPoolTaskScheduler::getActiveCount)
            .register(registry);
    FunctionCounter
            .builder("task.scheduler.pool-size", taskScheduler,
                    ThreadPoolTaskScheduler::getPoolSize)
            .register(registry);
}
}
```

现在，在 LibraryApplication 类上添加@EnableScheduling 注解并重新启动应用程序，Spring Boot 将收集并展现 TaskScheduler 的性能指标和健康状况数据，如图 10-7 所示。

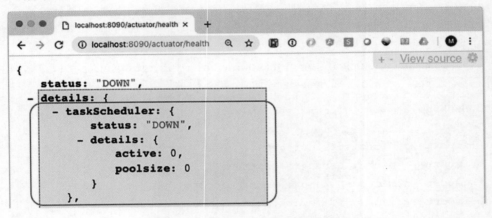

图 10-7　TaskScheduler 的健康状况检查

10.3　导出性能指标

10.3.1　问题

你希望将性能指标数据导出到外部系统，以创建一个仪表板来监视应用程序。

10.3.2　解决方案

使用一个 Spring Boot 支持的系统(如 Graphite)，并定期将性能指标数据推送到该系统。在应用程序中紧接着 spring-boot-starter-actuator 依赖项添加 micrometer.io 注册表的依赖项,将自动导出性能指标。默认情况下,Spring Boot 每分钟都会将数据推送到指定的服务器。

10.3.3 工作原理

导出性能指标是 Micrometer.io 库的一部分功能,它支持各种服务,如 Graphite、DataDog、Ganglia 或常规 StatsD。本节使用 Graphite,因此需要添加 micrometer-registry-graphite 依赖项。

```
<dependency>
  <groupId>io.micrometer</groupId>
  <artifactId>micrometer-registry-graphite</artifactId>
</dependency>
```

理论上,如果 Graphite(https://graphiteapp.org)位于本地主机,并且使用默认端口运行,那么添加上述依赖项之后就能够把性能指标数据发布到 Graphite 上了。但是,通过公开一些属性,Spring Boot 可以很容易地配置运行 Graphite 的服务器和端口号,这些属性通常包含在 management.metrics.export.<registry-name>名称空间中,如表 10-3 所示。

表 10-3 常用性能指标导出属性

属性	说明
management.metrics.export.<registry-name>.enabled	是否启用性能指标导出特性。当在类路径上检测到 Micrometer.io 库时默认为 true
management.metrics.export.<registry-name>.host	接收性能指标的主机,通常是本地主机或知名的服务网址(如 SignalFX、DataDog 等)
management.metrics.export.<registry-name>.port	接收性能指标的端口,默认为所需服务的已知端口
management.metrics.export.<registry-name>.step	发送性能指标数据的时间间隔,默认为 1 分钟
management.metrics.export.<registry-name>.rate-units	用于报告速率的基本时间单位,默认秒
management.metrics.export.<registry-name>.duration-units	用于报告持续时间的基本时间单位,默认毫秒

如果希望每隔 10 秒(而不是每分钟)报告一次性能指标,则需要将以下配置添加到 application.properties 文件中:

```
management.metrics.export.graphite.step=10s
```

注意：
bin目录下包含了graphite.sh文件，它使用Docker启动一个Graphite实例。

现在，这些指标将每 10 秒向 Graphite 发布一次。如果启动应用程序并在 http://localhost 上打开 Graphite(假设你正在运行前面提到的 Docker 容器)，那么可以创建一个 CPU 使用率的图表，如图 10-8 所示。

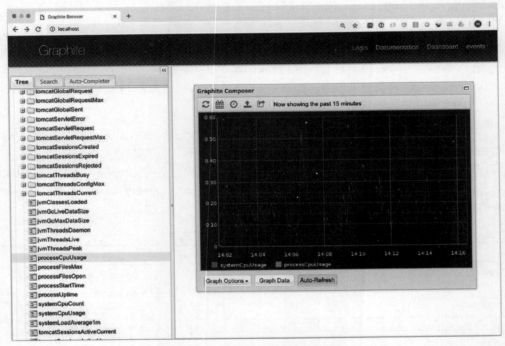

图 10-8　Graphite 上的 CPU 使用率

第 11 章

打 包

在本章中，你将了解对基于 Spring Boot 的应用程序进行打包的不同解决方案。

11.1 创建可执行文件

默认情况下，Spring Boot 会为应用程序创建一个 JAR 或 WAR 文件，这个文件可以通过命令 java -jar your-application.jar 运行。但是，你可能希望将应用程序作为服务器(目前已测试并支持基于 Debian 和 Ubuntu 的系统)启动的一部分运行。为此，你可以使用 Maven 或 Gradle 插件创建可执行 JAR。

11.1.1 问题

你需要一个可执行的 JAR，以便可以将其作为服务安装在你的环境中。

11.1.2 解决方案

Spring Boot 的 Maven 和 Gradle 插件都可以选择创建可执行的工件[1]。此时，归档文件也会变得、或者说表现得像启动/停止服务的 Unix Shell 脚本。

11.1.3 工作原理

要使工件可执行，需要按如下方式配置 Maven 或 Gradle 插件。

1. 使归档文件可执行

```
<plugin>
  <groupId>org.springframework.boot</groupId>
  <artifactId>spring-boot-maven-plugin</artifactId>
  <configuration>
    <executable>true</executable>
```

---

[1] https://docs.spring.io/spring-boot/docs/current/reference/html/deployment-install.html.

```
</configuration>
</plugin>
```

在 Maven 插件中,将 executable 属性设置为 true 后,生成的归档文件将是可执行的。

```
bootJar {
    launchScript()
}
```

现在,在构建工件之后,工件本身是可执行的,可以用来启动应用程序。此时可以简单地运行命令./your-application.jar 就能启动应用程序,而不需要使用命令 java -jar your-application.jar。

■ 提示:

有时可能需要使归档文件变成可执行的,可以使用 chmod +x your-application.jar 命令进行配置。

当使归档文件变得可执行时,归档文件将以 bash 脚本作为文件的前缀。你可能会认为这样做会破坏 Java 归档文件。但是,由于 Java 读取文件的方式(从下到上)和 shell 读取文件的方式(从上到下)不同,在归档文件中添加 bash 脚本实际上是没问题的。

可以使用命令 head -n 290 your-application.jar 查看归档文件中的 bash 脚本。

```
#!/bin/bash
#
#    .   ____          _            __ _ _
#   /\\ / ___'_ __ _ _(_)_ __  __ _ \ \ \ \
#  ( ( )\___ | '_ | '_| | '_ \/ _` | \ \ \ \
#   \\/  ___)| |_)| | | | | || (_| |  ) ) ) )
#    '  |____| .__|_| |_|_| |_\__, | / / / /
#   =========|_|==============|___/=/_/_/_/
#  :: Spring Boot Startup Script ::
#

### BEGIN INIT INFO
# Provides:          recipe_14_1
# Required-Start:    $remote_fs $syslog $network
# Required-Stop:     $remote_fs $syslog $network
# Default-Start:     2 3 4 5
# Default-Stop:      0 1 6
```

```
# Short-Description:    recipe_14_1
# Description:          Demo project for Spring Boot
# chkconfig:            2345 99 01
### END INIT INFO
```

使用 Maven 创建 Java 归档文件时，会将上述代码作为脚本的前部。这些描述信息将按照 Maven 给定的项目描述填充进来。你可以通过在 executable 属性之后指定 initInfoDescription 属性来覆盖这些描述。

2. 指定配置

通常，当你启动一个基于 Spring Boot 的应用程序时，有几个选项可以提供额外的配置(参见 2.2 小节)。但是，当将归档文件用作脚本(或作为服务)时，其中一些方式将不再适用。相反，你可以在紧接着可执行文件的位置添加.conf 文件(必须将其命名为 your-application.conf)来包含应用程序的其他配置选项，可用的选项如表 11-1 所示。

表 11-1 可用的属性

属性	描述
MODE	操作的"模式"。默认值是 auto，它将检测模式；从 symlink 启动时，其行为与 service 类似。如果要在前台运行进程，请更改为"run"
USE_START_STOP_DAEMON	是否使用 start-stop-daemon 命令。默认情况下，将检测命令是否可用
PID_FOLDER	写入 PID 的文件夹名称，默认为/var/run
LOG_FOLDER	写入日志的文件夹名称，默认为/var/log
CONF_FOLDER	读取.conf 文件的文件夹名称，默认与 jar 文件的目录相同
LOG_FILENAME	写入日志的文件名称，默认值为<appname>.log
APP_NAME	应用程序的名称。如果 jar 是从 symlink 运行的，脚本会猜测应用程序的名称
RUN_ARGS	传递给 Spring Boot 应用程序的参数
JAVA_HOME	默认情况下，将从$PATH 中查找；如果需要，可以显式定义
JAVA_OPTS	要传递给 JVM 的选项(如内存设置、GC 设置等)
JARFILE	JAR 文件的显式位置，以防脚本用于启动实际上未嵌入的 JAR
DEBUG	如果不为空，则在 shell 进程上设置-x 标志，以便于查看脚本中的逻辑
STOP_WAIT_TIME	强制关机前等待的时间(默认为 60 秒)

当使用嵌入式脚本(默认的行为)时，不能在.conf 文件中配置 JARFILE 和 APP_NAME 属性。

```
JAVA_OPTS=-Xmx1024m
DEBUG=true
```

在 .conf 文件中添加上述配置后，Spring Boot 将为应用程序分配最大 1GB 的内存，同时应用程序将为 shell 脚本做一些额外的日志记录。

11.2 为部署创建 WAR 文件

11.2.1 问题

你需要创建的不是 JAR 文件，而是用于部署到 Servlet 容器或 JEE 容器的 WAR 文件。

11.2.2 解决方案

将应用程序从 JAR 打包为 WAR，并让 Spring Boot 应用程序扩展 SpringBootServletInitializer 类，以便它可以作为常规应用程序引导自己。

11.2.3 工作原理

要使 3.1 小节的应用程序成为可部署的 WAR 文件，需要做三件事：
(1) 将打包选项从 JAR 修改为 WAR。
(2) 扩展 SpringBootServletInitializer 类以在部署时引导应用程序。
(3) 将嵌入式服务器的作用域更改为 provided。
在 pom.xml 中，将打包选项从 JAR 更改为 WAR。

```xml
<packaging>war</packaging>
```

为了防止嵌入式容器向 Web 应用添加它的类，需要将嵌入式容器的作用域更改为 provided，而不是默认值 compile。
使用嵌入式 Tomcat(默认)时，将以下内容添加到 pom.xml 文件中。

```xml
<dependency>
  <groupId>org.springframework.boot</groupId>
  <artifactId>spring-boot-starter-tomcat</artifactId>
  <scope>provided</scope>
</dependency>
```

当使用不同的嵌入式容器时(参见 3.7 小节)，通过添加<scope>provided</scope>将容器的范围修改为 provided。这样做时，Spring Boot 插件不会将相关的库文件添加到默认的 WEB-INF/lib 目录中。但是，这些库文件仍然是所创建的 WAR 文件的一部分，只是位于特殊的 WEB-INF/lib-provided 目录中。Spring Boot 知道这个位置，因此也可以使用 WAR 来启动嵌入式容器。启动嵌入式容器对于开发是非常有利的，但由于它是一个 WAR 文件，因此它也可以部署到容器中，比如 WebSphere 的 Tomcat 容器。

为了确保应用程序能够启动，需要应用程序扩展 SpringBootServletInitializer 类。这是在 Servlet 或 JEE 容器中引导 Spring Boot 所需的特殊类。

```java
package com.apress.springbootrecipes.helloworld;

import org.springframework.boot.SpringApplication;
import org.springframework.boot.autoconfigure.SpringBootApplication;
import org.springframework.boot.builder.SpringApplicationBuilder;
import org.springframework.boot.web.servlet.support.SpringBootServletInitializer;

@SpringBootApplication
public class HelloWorldApplication extends SpringBootServletInitializer {
  public static void main(String[] args) {
      SpringApplication.run(HelloWorldApplication.class, args);
  }

  @Override
  protected SpringApplicationBuilder configure(SpringApplicationBuilder builder) {
      return builder.sources(HelloWorldApplication.class);
  }
}
```

在扩展 SpringBootServletInitializer 类时，需要覆盖 configure 方法。configure 方法获取一个 SpringApplicationBuilder 实例，可以使用它来配置应用程序。要添加的内容之一是主配置类，就像 SpringApplication.run 一样。代码 builder.sources(HelloWorldApplication.class)完成上述动作。这将用于引导应用程序。

为了完整起见，下面是在 3.7 小节中编写的 HelloWorldController 类。

```java
package com.apress.springbootrecipes.helloworld;

import org.springframework.web.bind.annotation.GetMapping;
import org.springframework.web.bind.annotation.RestController;

@RestController
```

```
public class HelloWorldController {

  @GetMapping
  public String hello() {
    return "Hello World, from Spring Boot 2!";
  }
}
```

■ 警告：
将某些内容部署到Servlet或JEE容器时，Spring Boot不再受服务器控制。由于此配置，server和management.servlet名称空间中的配置选项不再适用。因此，如果定义了server.port，则在部署到外部服务器时将忽略它！

部署应用程序时，它在根目录/上不再像使用嵌入式服务器时那样可用。然而，可以通过/<name-of-war>/访问应用程序：通常，通过类似 http://<name-of-server>:8080/<name-of-war>的地址访问应用程序。当部署到一个标准的 Tomcat 安装并查看Management GUI 时，界面类似于图 11-1 所示。

Applications		
Path	Version	Display Name
/	None specified	Welcome to Tomcat
/docs	None specified	Tomcat Documentation
/examples	None specified	Servlet and JSP Examples
/host-manager	None specified	Tomcat Host Manager Application
/manager	None specified	Tomcat Manager Application
/spring-mvc-as-war-2.0.0	None specified	

图 11-1　Tomcat 管理图形用户界面

点击/spring-mvc-as-war-2.0.0 链接将打开应用程序，如图 11-2 所示。

图 11-2　已部署的应用程序的运行结果

11.3　通过 Thin Launcher 减少归档文件大小

11.3.1　问题

默认情况下，Spring Boot 会生成一个所谓的胖 JAR，一个包含所有依赖项的 JAR。这种打包方式有一些明显的好处，因为 JAR 是完全自包含的。但是，JAR 的大小会显著增加。当运行多个应用程序时，你可能希望重用已经下载的依赖项，以减少整体占用的空间。

11.3.2　解决方案

打包应用程序时，可以指定自定义布局。其中一个自定义布局是 [Thin Launcher][2]。这个 Launcher 将使得应用程序在启动之前下载依赖项，而不是使用在应用程序内打包的依赖项。

11.3.3　工作原理

按原样使用 Spring Boot 插件时，所有需要的库都包含在归档文件的 BOOT-INF/lib 文件夹中。然而，人们可能希望减小 JAR 的大小，并允许重用依赖项，从而创建总体占用空间更小的归档文件。例如，14.1 小节编写的应用程序生成的归档文件大约为 7.1M，其中大部分来自所包含的依赖项。

可以为 Spring Boot 插件添加自定义的布局和启动器。布局定义了从何处加载源，启动器使用这些信息来加载对应的依赖项。Thin Launcher 将从 Maven Central 下载这些工件，并将它们放置在一个共享的存储库中。因此，当多个应用程序共享依赖项时，它们只会被下载一次。

要使用 Thin Launcher，需要添加 Spring Boot 插件的依赖项，如下所示。

---

2 https://github.com/dsyer/spring-boot-thin-launcher.

```xml
<plugin>
  <groupId>org.springframework.boot</groupId>
  <artifactId>spring-boot-maven-plugin</artifactId>
  <dependencies>
    <dependency>
        <groupId>org.springframework.boot.experimental</groupId>
        <artifactId>spring-boot-thin-layout</artifactId>
        <version>1.0.15.RELEASE</version>
    </dependency>
  </dependencies>
</plugin>
```

这足以创建一个相当小的归档文件。在运行命令./mvn package 时，生成的 JAR 大约为 12K。但是，缺点是在启动应用程序时需要下载依赖项，因此根据依赖项的数量，启动应用程序可能需要更长的时间。

添加插件的依赖项之后，Spring Boot 将为应用程序添加一个名为 ThinJarWrapper 的类，该类将成为应用程序的入口点。现在，应用程序包含 pom.xml 文件和/或 META-INF/thin.properties 来确定依赖项。thinJarWrapper 将找到另一个 JAR 文件(即"启动器")。打包器下载启动器(如果需要)，或者使用本地 Maven 存储库中的缓存版本。

然后，启动器接管启动流程并读取 pom.xml 文件(如果存在)和 META-INF/thin.properties 文件，并根据需要下载依赖项(和所有传递项)。现在，它将创建一个自定义的类加载器，所有已下载的依赖项都保存在类路径上。然后用这个类加载器运行应用程序自己的入口方法。

■ 提示：
来自Maven的依赖项下载完成后，将沿用在Maven settings.xml中做出的设置，因为它使用常规的Maven工具。当为Maven存储库使用一个类似Nexus的镜像时，需要在META-INF/thin.properties文件中包含thin.repo属性来指向那个镜像，否则下载启动器就会失败。

11.4 使用 Docker

使用 Docker[3] 构建和发布容器是现在的普遍做法。当使用 Spring Boot 时，很容易将容器包装到一个 Docker 容器中。

---

[3] https://www.docker.com.

11.4.1 问题

你希望在 Docker 容器中运行基于 Spring Boot 的应用程序。

11.4.2 解决方案

创建一个 Dockerfile 并使用一个可用的 Maven Docker 插件来构建 Docker 容器。

> ■ **注意：**
> 在本节中，我们选择使用 Spotify 中的 dockerfile-maven-plugin 插件，但还有其他插件同样有效。

11.4.3 工作原理

首先，需要一个包含构建容器所需信息的 Dockerfile。当容器构建完毕后，你可以使用 docker run 命令启动它。

1. 更新构建脚本以生成 Docker 容器

要创建 Docker 容器，首先创建 Dockerfile 文件。Dockerfile 是一个文件，它包含了如何构建容器的信息。将这个文件放在项目的根目录中。

```
FROM openjdk:11-jre-slim
VOLUME /tmp
ARG JAR_FILE
COPY ${JAR_FILE} app.jar
ENTRYPOINT ["java","-Djava.security.egd=file:/dev/./urandom",
"-jar","/app.jar"]
```

通过上述 Dockerfile 文件，我们使用 openjdk 提供的容器来构建我们自己的容器。通过 ADD 命令将我们的应用程序添加到该容器，最后告知该容器在启动期间要启动的内容，为此可以使用 ENTRYPOINT 命令。

> ■ **警告：**
> 使用公开可用的容器作为起点似乎是一个好主意，但你必须意识到这可能带来的安全隐患。由于无法控制特定的容器，因此无法保证容器中内置(或不内置)的内容。对于实际情况，你可能希望构建自己的基础容器。

带 JAR_FILE 参数的 ARG 命令告诉此构建有一个可在构建脚本中使用的变量 JAR_FILE，我们将通过插件配置提供该变量的值。

ENTRYPOINT 仅仅指定了它将运行 java -jar/app.jar 命令。你还可以将其与 14.1

小节结合起来，安装一个可执行的 JAR 作为脚本，从而使入口占用的空间更小一点。

现在编辑好了 Dockerfile，需要将 dockerfile-maven-plugin 添加到 pom.xml 的构建部分中。

```xml
<plugin>
  <groupId>com.spotify</groupId>
  <artifactId>dockerfile-maven-plugin</artifactId>
  <version>1.4.4</version>
  <configuration>
    <repository>spring-boot-recipes/${project.name}</repository>
    <tag>${project.version}</tag>
    <buildArgs>
      <JAR_FILE>target/${project.build.finalName}.jar</JAR_FILE>
    </buildArgs>
  </configuration>
  <dependencies>
    <dependency>
      <groupId>javax.activation</groupId>
      <artifactId>javax.activation-api</artifactId>
      <version>1.2.0</version>
    </dependency>
    <dependency>
      <groupId>org.codehaus.plexus</groupId>
      <artifactId>plexus-archiver</artifactId>
      <version>3.6.0</version>
    </dependency>
  </dependencies>
</plugin>
```

我们通过插件的 repository 属性指定映像的名称，这样就知道以后在哪里可以找到容器。通常，你还希望在容器上放置一个标记，这里我们选择使用 ${project.version} 作为标记。

buildArgs 用于向 docker build 命令传递参数，以便在处理 Dockerfile 时可以使用这些参数。由于已经指定了 JAR_FILE 参数，所以我们还在配置中声明了这一点，并将其指向生成工件的位置。

■ 注意：
指定 javax.activation-api 依赖项的原因是它不是 Java 9 可用类默认集合的一部分；如果你使用的是 Java 8，则不需要这样做。指定 plexus-archiver 依赖项的原因是

Maven 的默认版本在 Java 9 及以上的版本中不起作用。

2. 构建和启动容器

现在一切就绪，你可以使用 mvn clean build dockerfile:build 命令来生成 Docker 容器。

要启动容器，可以使用如下命令运行它：

```
docker run -d spring-boot-recipes/dockerize:2.0.0-SNAPSHOT
```

这将在后台启动容器，并开始将消息打印到控制台。要查看日志，请执行命令 docker logs <name-of-container> --follow，你将看到每两秒钟打印一条消息，如图 11-3 所示。

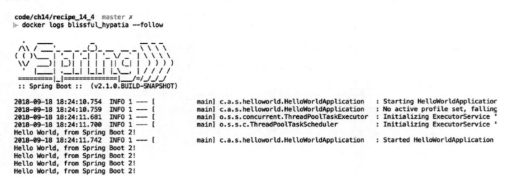

图 11-3　Docker 容器日志

3. 将属性传递给 Spring Boot 应用程序

在 Docker 容器中运行 Spring Boot 应用程序时，你可能希望能够根据应用程序部署的环境更改属性。为此可以使用 Spring 配置文件(请参见第 2 章中的内容)，但你仍然需要提供一个变量，指定要使用哪个配置文件。一般情况下，可以使用类似 java-jar your-application.jar --spring.profiles.active=profile1,profile 的命令启动应用程序。但是，由于使用了 Docker 容器，这是不可能的。

幸运的是，使用 Docker 容器时，可以通过-e 开关传递环境变量，而且由于 Spring Boot 还考虑了本地环境中的变量[4]，所以可以使用-e 开关。

对于这里编写的简单应用程序，可以传递 audience 属性(默认情况下，它将发送 World，服务器收到消息后返回消息 Hello World, from Spring Boot 2! 作为响应)。让我们把发送的内容换成 Docker。为此，需要将-e AUDIENCE='Docker'添加到 run 命令中。

```
docker run -d -e AUDIENCE='Docker'
```

---

4 https://docs.spring.io/spring-boot/docs/current/reference/html/boot-featuresexternal-config.html。

```
spring-boot-recipes/dockersize:2.0.0-SNAPSHOT
```

在查看重新启动的容器的日志记录时，消息已经更改为 Hello Docker, from Spring Boot 2!，如图 11-4 所示。

```
  '  |____| .__|_| |_|_| |_\__, | / / / /
 =========|_|==============|___/=/_/_/_/
 :: Spring Boot ::  (v2.1.0.BUILD-SNAPSHOT)

2018-09-18 18:27:00.941  INFO 1 --- [           main] c.a.s.helloworld.HelloWorldApplication   : Starting HelloWorldApplicatior
2018-09-18 18:27:00.948  INFO 1 --- [           main] c.a.s.helloworld.HelloWorldApplication   : No active profile set, fallinc
2018-09-18 18:27:01.863  INFO 1 --- [           main] o.s.s.concurrent.ThreadPoolTaskExecutor  : Initializing ExecutorService '
2018-09-18 18:27:01.882  INFO 1 --- [           main] o.s.s.c.ThreadPoolTaskScheduler          : Initializing ExecutorService '
2018-09-18 18:27:01.916  INFO 1 --- [           main] c.a.s.helloworld.HelloWorldApplication   : Started HelloWorldApplication
Hello Docker, from Spring Boot 2!
Hello Docker, from Spring Boot 2!
Hello Docker, from Spring Boot 2!
Hello Docker, from Spring Boot 2!
Hello Docker, from Spring Boot 2!
```

图 11-4　Docker 容器的输出

■ **注意：**

当在基于Unix的系统中设置变量时，通常需要将所有变量名修改为大写字母，并将所有的 . 替换为 _。例如，在传入spring.profiles.active时，需要修改为SPRING_PROFILES_ACTIVE。
